T0074562

Molds, Mushrooms, and Medicines

Molds, Mushrooms, and Medicines

Our Lifelong Relationship with Fungi

Nicholas P. Money

PRINCETON UNIVERSITY PRESS

PRINCETON & OXFORD

Published by Princeton University Press
41 William Street, Princeton, New Jersey 08540
99 Banbury Road, Oxford OX2 6JX

press.princeton.edu

All Rights Reserved

Library of Congress Cataloging-in-Publication Data

Names: Money, Nicholas P., author.
Title: Molds, mushrooms, and medicines : our lifelong relationship with fungi / Nicholas P. Money.
Description: Princeton : Princeton University Press, [2024] | Includes bibliographical references and index.
Identifiers: LCCN 2023030612 (print) | LCCN 2023030613 (ebook) | ISBN 9780691236308 (hardback) | ISBN 9780691236315 (ebook)
Subjects: LCSH: Fungi. | Materia medica, Vegetable. | Molds (Fungi). | BISAC: SCIENCE / Life Sciences / Mycology | NATURE / Plants / Mushrooms
Classification: LCC QK603 .M58 2024 (print) | LCC QK603 (ebook) | DDC 579.5—dc23/eng/20230908
LC record available at https://lccn.loc.gov/2023030612
LC ebook record available at https://lccn.loc.gov/2023030613

British Library Cataloging-in-Publication Data is available

Editorial: Alison Kalett, Hallie Schaeffer
Jacket: Heather Hansen
Production: Jacqueline Poirier
Publicity: Matthew Taylor (US), Kate Farquhar-Thomson (UK)
Copyeditor: Susan Campbell

Jacket Credit: Jacket images (clockwise): Nataliya Hora / Alamy Stock Photo; Eye of Science / Science Source; BSIP SA / Alamy Stock Photo; Guido Blokker / Unsplash

This book has been composed in Arno Pro and URW DIN Arabic.

Printed in the United States of America

10 9 8 7 6 5 4 3 2 1

Contents

Acknowledgments

I WOULD LIKE to thank my agent, Deborah Grosvenor, and my editor, Alison Kalett, for making this book happen. Andor Kiss helped me to untangle the studies on ghost gut fungi, Michael Klabunde assisted with Latinisms, and my wife, Diana Davis, proofread the developing manuscript.

1

Interacting

ENCOUNTERS WITH FUNGI FROM BIRTH TO DEATH

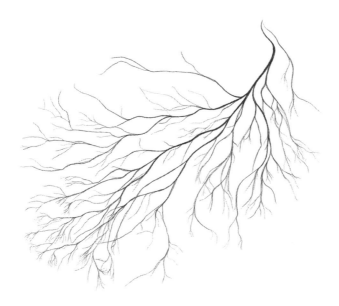

Sing, heavenly Muse . . .
What in them is dark
Illumine, what is low raise and support;
That to the height of this great argument
I may assert eternal providence,
And justify the ways of mushrooms to men.

—JOHN MILTON, *PARADISE LOST* (1667),
BOOK I, LINES 6, 22–26, AMENDED

FUNGAL SPORES cast a shadow over my childhood and almost killed me. One day in 1967, my five-year-old body began to run out of oxygen as my lungs shut down with inflammation, turning my skin blue before the ambulance arrived. I was born in the Thames Valley of southern England, which is a lovely place unless you are an asthmatic. Tree pollen and fungal spores fill the Oxfordshire air in summer and turn paradise into hell. There had been a thunderstorm that afternoon, which whipped clouds of these noxious particles into the sky. They filtered into my chest with each breath, causing my little airways to narrow and flood with mucus. A severe asthma attack feels like death. The nurses put me in an oxygen tent and gave me big orange tablets that were difficult to swallow, but after a day or two these antibiotics combined with a steroidal medicine reopened my lungs. More than fifty years later I can still see myself struggling to breathe under that clear plastic canopy, and I wonder how much this trauma led to my career as a mycologist and immersion in research on the spores of the fungi that put me there.

The fact that a boy plagued by spores became a scientist who has spent his adult life studying fungi, teaching students about their biology, and serving as an expert witness in lawsuits related to mold exposure is one of those serendipitous outcomes that define so many lives. The connections between my childhood and my profession did not occur to me until my brief experience as the patient of a therapist. He was a gentle, bearded man who asked insightful questions as he sought to help me understand why I was wrestling with thanatophobia, or death anxiety, which was distracting me from enjoying not being dead. Early in our conversations, I told him that I was an expert on fungi, a mycologist, then explained a little about what a fungus is and what a fungus does. We talked about many other things—my health, marriage, and the challenges of parenting teenagers—before he circled back one day and asked: "Have you ever wondered why you are obsessed with death and with the microbes that you have described as the great decomposers?" We both laughed. It seems plausible that my asthma attacks were the foundation of it all: thanatophobia and what some would view as a morbid fascination with fungi that may have developed as a subliminal attempt at therapy, like the hypochondriac who becomes a doctor. On the other hand, maybe I just liked mushrooms.

What intrigues me now, and is the subject of this book, is the science of the human-fungus symbiosis, both the intimate and the extended relationship between fungi and our species. This relationship runs all the way from yeasts that grow on the skin and inside the gut to our uses of fungi as food and sources of medicines and, ultimately, to the mushroom colonies in soil that make life on land possible. Our closest physical ties with the fungi are invisible because the ones that live on the body are microscopic. These species grow amid the more numerous bacteria and viruses and are critical players in human health. Together, these microbes form the human microbiome, and we identify the fungal part of this intimate ecosystem as the *mycobiome*. (The prefix, *myco-*, from the Greek *mykes*, refers to all things fungal.)

The growth of immense numbers of fungi on the skin and inside the body is an unexpected and startling fact of science. Fungi are a vital part of the immense ecosystem of the human body, which operates as a partnership between trillions of human and microbial cells. We cannot live without these fungi. Touch the creased skin behind one of your ears or run your hands through your hair. You will not see them, of course, but fungal cells will cling to your fingertips afterward and every other time you rub, scratch, pick, or caress. They are essential partners, lodgers on all of us. Most of the fungi of the mycobiome are helpful, but some can turn on us when our immune defenses are weakened and cause terrifically damaging infections. Fungi that normally grow on plants, rotting wood, compost heaps, and bird droppings can also settle on the body and attack our tissues if we are vulnerable. Human diseases caused by fungi are called *mycoses*, and these range from the irritation of athlete's foot to life-threatening brain infections.

But our relationship with fungi does not end with the species found on the body. It widens to our conscious interactions with these microbes through their roles in our diet and as a source of powerful medicines. Science has been advancing in all of these areas of mycological inquiry, from studies that reveal the diversity of yeasts that grow on the skin to research on the use of psychedelic mushrooms in the treatment of depression. Once we expand our view of the give-and-take between humans and fungi to these deliberate uses of fungi, we discover a broader relationship, a human-fungus symbiosis that is a defining feature of our biology and culture. The term "symbiosis" is used in its original and most liberal sense in this book to describe helpful and harmful relationships between species. This is a perfect reflection of the incredible range of interactions between humans and fungi.

WHAT IS A FUNGUS?

Not plant, not animal, more animal than plant, and treated as the most mysterious kingdom of life in popular culture, fungi come in many shapes and sizes.[1] The fungi we see most often seem too big to be categorized as microbes. These are mushrooms, which include the fairy-tale fly agaric, with its red cap spotted with white scales; shelf fungi, as big as dinner plates, that grow on decomposing logs; and slices of white button mushrooms on pizzas. The reason we call these species micro-organisms is that the fungus that forms the mushroom *is* microscopic. For almost all of their lives, these organisms exist as spidery colonies of tiny threads called hyphae. Each thread, or hypha, is ten times thinner than a human hair. These filaments elongate and branch as they feed in soil and go about the process of rotting wood. The colony of branching hyphae is a mycelium. When this mycelium has grown over a large area and absorbed enough food, it reverses direction and flows to the surface, where the threads merge to form mushrooms. Mushrooms with gills are the fruit bodies or sex organs of fungi that mist the air with spores. As the urge to reproduce becomes an imperative, the fungus moves from belowground to aboveground, changing its role from feeding to fruiting in the wondrous cycle of its life.

But most fungi never form a mushroom and are microscopic throughout their feeding and reproductive stages. These include aquatic fungi that swim in ponds, with tailed cells that resemble animal sperm; molds with stalks hung with sparkling spores that look like miniature chandeliers; and 1,500 species of yeasts. Yeasts include the species used in brewing and baking, whose Latin name is *Saccharomyces cerevisiae*, and another fungus, called *Candida albicans*, that lives on everyone and is best known, unfortunately, for its irritating nature as the vaginal yeast.[2] (Latin names are kept to a bare minimum in this book, but some of the fungi are best known through their Latin names, and others are so obscure that they have never been given a common name.) Unlike fungi that grow as thin threads, which we call molds, yeasts develop as single rounded cells and produce buds, or daughter cells, on their surface.

MAKING SENSE OF THE MYCOBIOME

The entire human body is affected by fungi. Yeasts populate the skin and crowd around the hair follicles on the scalp; other species live in the ear canals, nasal passages, and mouth; and fungi swarm in the digestive and reproductive systems. The fungi are as small as the cells of our tissues and only become visible when they grow in such profusion that they form patches and pastes on the surface of the skin. But at this moment, and throughout our lives, fungal cells are feeding on the scalp and growing in the gut, consuming the mucus and dead cells that we discharge every day, and helping to control the bacteria. This microbial community is in constant flux—and so much of it was completely unknown until recently.

Fungi have been sidelined at meetings of microbiologists by studies on bacteria and viruses during my career and have been an afterthought in medicine. Earlier generations of mycologists misunderstood the fungi that they found on the body, regarding most of them as germs that damaged hospital patients and overlooking the significance of the yeasts growing peacefully on everyone else. Even when molecular genetic techniques began to reveal the incredible diversity and number of microbes in the gut, the fungi were missed because the methods were limited to identifying the DNA sequences of bacteria. This picture is changing at last, and new investigative methods are exposing the yeasts and molds—the blobs and filaments—multiplying from scalp to toes on the outside of the body and from mouth to anus on the inside. As this examination of the fungi has proceeded, the vision of the microbiome as a mostly bacterial territory has shifted to an appreciation of the diverse communities of fungi that fight and cooperate with bacteria through webs of chemical interactions to make a living on the body.[3] Through these innovations we are beginning to fathom the extraordinary influence of the mycobiome on our health and well-being.

Even with this knowledge, the invisibility of the human mycobiome makes it difficult to comprehend. This is life-changing science in the sense that it permits a new view of the body, but it takes some imagination too. There is nothing cellular and microbiological about the way we look and feel. We picture ourselves as semisolid individuals, singular beings constructed with varying degrees of smoothness and raggedness, pert here, drooping there, and hanging together across the decades, but this half-truth belies our biological nature. For a more enlightened sense of self, we have to close our eyes to picture the body as a galaxy of cells, to say, and to believe, "I am a trinity, born from one cell, copied in trillions, and filled with other forms of life."

Microbes that grow naturally on the body belong to the healthy microbiome. Different microbes populate different locations on the body, including the skin, ear canals, nasal passages, lungs, teeth, digestive tract, and reproductive system. In each place, the body supports intermingled communities of bacteria, viruses, and fungi that form the bacteriome, virome, and mycobiome. The term mycobiome was used first to describe the fungi associated with plants in a salt marsh on the Eastern Shore of Virginia and now applies to the communities of fungi found in any location.[4] For example, cultivated pineapples harbor forty-nine species of fungi, tropical corals host a jumble of marine fungi, and the gut of the largest living lemur, called the indri, is heaving with microbes acquired from its vegan diet and supplemented with fungi from the soil.[5] Every animal, plant, and seaweed is crawling with fungi.

We see hints of the ancestral human mycobiome in monkeys and other apes, whose gut fungi vary among species.[6] The community of fungi that is distinctively human has been remodeled from earlier mixtures of the fungi that grew on our hominid ancestors. This has involved losses and replacements of fungi as the mycobiome has adapted to changes in our diet and behavior. These evolutionary modifications have unfolded over the course of millions of years, but fungi have also come and gone on much shorter historical timescales. The communities of fungi on the body were rejiggered when we emigrated from our birthplace in Africa to other climatic zones and began fabricating clothing and footwear. Clothes and shoes affected the fungi on the skin, and the gut fungi were modified as we exchanged hunting and gathering for life in agricultural settlements.

WHAT THE FUNGI ARE DOING ON THE BODY

Fungi grow on humans because fungi grow everywhere they find food, and we are stuffed with energy. A pound of flesh has the same number of calories as a pound of ice cream, for an average of more than one hundred thousand calories per body.[7] Colonies of yeasts consume a snippet of this when they digest the natural oils on the scalp. Other fungi live by breaking down bacteria and food particles as they are squeezed along the 1.5-meter-long tube of the colon. We are the unconscious hosts of a 24/7 buffet.

Although we are unaware of these fungi unless they cause inflammation and tissue damage, the immune system is policing them throughout the day. This defense mechanism is a marvel of natural engineering. White blood cells are the biggest players in immunology. These colorless cells flow in the blood alongside the more plentiful red blood cells. Red blood cells outnumber white blood cells by six hundred to one, and all of these red blood cells do exactly the same thing: they pick up oxygen from the fresh air entering the lungs when we inhale and carry it around the body; on the return path, the red blood cells release the carbon dioxide that we exhale. White blood cells have nothing to do with this import and export of gases and are more diverse in their roles. Following their own genetic programming, white blood cells called phagocytes move around the body, stretching themselves forward and retracting behind, just like amoebas that live in soil and water. Some of them spread from the bloodstream into the surrounding tissues and crawl into the linings of the lungs and the outer layers of the skin. Wherever they encounter microorganisms, they decide whether to kill them or give them a pass. White blood cells also rid the body of human cells that are damaged or start to grow in a way that can lead to the development of tumors. We would not last for very long at all without these defenders of the immune system.

The immune system manages the whole ecosystem of human tissues and embedded microorganisms. It allows specialized fungi to feed on the dead cells on the skin surface and the fatty secretions on the scalp and other fungi to multiply in the mucus linings of the digestive and respiratory systems. Turning to its more aggressive role, the immune system works continuously to eliminate insurgents, including the fungal pathogens that threaten to choke the whole enterprise. The resulting mixtures of fungi that live on an individual body are unique.[8] The types of fungi differ in each location, and their numbers change as some grow and reproduce and others deteriorate and die. Our age and gender affect the numbers and kinds of microbes on the body too, and there is some evidence that hormonal differences between the sexes may stimulate the growth of distinctive groups of fungi. Sweating stimulates the skin residents to proliferate, sun exposure kills others; gardeners pick up fungi from the soil and from plants; children transfer fungi to their parents and vice versa; and lovers swap fungi in bed. The number and identity of the spores flowing through the nose and into the lungs vary as we breathe, and the fungi in the mouth and digestive system are agitated with each meal.

Geography is another variable, because fungi are not distributed evenly across the planet. This means that there is a regional character to the mycobiome, with people in Africa partnering with different fungi than Asian populations. Diet has a dominating effect on the kinds of fungi and their abundance in the gut, with major differences between vegetarians, meat eaters, and consumers of lots of processed foods.[9] Many illnesses affect the types and abundance of fungi, especially those in the gut, and traumatic injuries including severe skin burns enable fungi to penetrate deep into our tissues. The catalog of these sundry influences on the composition of the fungal communities on the body is endless.

The significance of human-fungal interactions is surprising, at first, when we consider the scarcity of the fungi compared with the bacteria in the microbiome. There are *only* 40 billion fungal cells in the human gut, compared with 40 trillion bacterial cells, a thousand bacteria to each fungus.[10] Heaped together, the bacteria weigh as much as a cup of sugar, and spread in a single line, they would encircle Earth. The less numerous but far larger fungal cells weigh no more than a raisin, but their combined surface area is impressive—equivalent to an eight-person dining table. The interior lining of the large intestine covers a similar area, and even though many of the fungi are buried in the dissolving food and developing feces, these calculations reveal that the fungi offer a lot of real estate for chemical interactions with the body. The molecule-by-molecule transactions between the immense surface presented by the fungi and the immune system go a long way toward explaining how such tiny organisms can have such a profound effect on our health.

The impact of the fungi on the health of the gut is a controversial subject among microbiome experts. Some view the fungi as critical players in the internal ecosystem, while others believe that their activities are eclipsed by the overwhelming number of bacteria.[11] These differing viewpoints have arisen because the science of the mycobiome is moving at lightning speed, and conclusions about the influence of the fungi on some health conditions swerve from study to study. But the truth is unfolding, and this book will share consensus views when they are available and highlight other areas of mycobiome research where we remain puzzled.

Uncertainties come from many sources. The first and most formidable of these is the question of cause versus effect.[12] If we find differences in the abundance of fungi in people suffering from a particular illness, it can be very difficult to determine whether the fungi are responsible for the illness or whether the change in their number is a consequence of another issue with the patient's health. Either way, treating the fungus may be effective at alleviating some of the most distressing symptoms of the complaint. Examples of these conditions that do not seem to be caused by fungi but are associated with shifts in the onboard populations of fungi are explored in the chapters that follow. The second diagnostic problem comes from the awesome power of the modern genetic techniques that allow us to detect the dead cells of fungi that are passing through the intestine and may have entered the body in our food. This makes it difficult to identify the living species that are long-term residents in the gut, which are more important in the health of the digestive system than the remains of dietary migrants.

While the relevance of the fungi in understanding some aspects of healthy gut function is debatable, a substantial body of research has proven that fungi belong to communities of microorganisms in other parts of the body where they are crucial in health and disease. There is no question about the importance of the fungi on the body surface, where they support skin health and can also cause allergic responses and unsightly infections. Similarly, we know that fungal spores are powerful triggers of asthma and other allergic illnesses in the respiratory system. Lastly, it is important to recognize the connections between the mycobiome in different parts of the body. The effects of the gut fungi on the immune system can influence the development of health conditions elsewhere, and fungi can also move physically from one location to another—from skin to gut and vice versa. With so much attention to the effects of fungi on human health, this is the perfect time to explore this phenomenal symbiosis.

WHEN THINGS GO WRONG

Everyone is damaged by fungi at some time in their lives. Hundreds of millions of people suffer from allergies caused by fungal spores, and fungal infections range from skin irritation and toenail disintegration to the foulest flesh-melting diseases ever pictured in a pathology textbook. The World Health Organization (WHO) prioritized the surveillance of serious fungal infections in 2022 with the publication of a list of nineteen species that pose the greatest threat to public health.[13] This was in response to the rising numbers of fungal infections and difficulties in diagnosing and treating these serious illnesses. Mycoses kill 1.5 million people every year, and the threat of the severest infections is growing with the emergence of strains of yeasts that are resistant to antifungal medicines. The burden of these illnesses is heightened in the developing world, where access to adequate medical care is limited, and in more affluent countries with aging populations that are more vulnerable to infection. Poisoning is another hazard of our relationship with the fungi, resulting from the mistaken identification of wild mushrooms by foragers and by consuming the toxins produced by molds that grow on harvested grain. These harmful interactions complete the picture of our relationship with the fungi, which runs from beauty to the beast, the yin and the yang of the human-fungus symbiosis.

It is easy to disregard the fungi that support our health because they are invisible, and we can live in blissful ignorance of the dark side of mycology if we are fortunate to avoid infections. But wisdom comes from familiarity with this subject. The types of fungi that live on the body and their levels of activity vary according to where we live, what we eat, and whether we work indoors or outdoors. Changes in our health, drugs that we are prescribed, and the use of consumer products including toothpaste, shampoos, and body lotions modify our closest relationship with fungi too. No amount of grooming will leave us unfungal, which is a good thing because we would be in bad shape if we abolished our partners. A body without fungi would be as barren as a forest without mushrooms. Equipped with this self-knowledge, we can seek opportunities for correcting imbalances in this intimate relationship.

BEYOND THE BODY: THE EXTENDED SYMBIOSIS

Beyond the body, fungi inhabit our pets, are active in damp places in our homes, and flourish on fruits and vegetables in the kitchen. Pet dogs and cats are covered with yeasts, bathrooms are mycological playgrounds, and we consume fungi clinging to salad ingredients without giving them a thought—until the tomatoes sprout hairs. Physical contact with these fungi is certain, continuous, and consequential. There is a historical character to these passive interactions with the fungi in the environment, which has changed over thousands of years and continues to shift in the modern world. The transition from hunter-gatherer and nomadic lifestyles to agricultural settlements around ten thousand years ago had a profound effect on our interactions with fungi, by exposing us to masses of fungal spores emanating from moldy grain stores and to toxins in food made from these spoiled cereals. Asthma and other allergies to fungal spores were born from changes in farming over the millennia, and the crowding of populations in cities promoted the spread of skin infections by ringworm fungi. The human-fungus relationship has also operated on a more conscious level with the incorporation of fungi into our diet, from foraging for wild mushrooms and cultivating an increasing selection of species, to the growing popularity of nuggets of fungal protein or mycoprotein manufactured in bioreactors. The uses of fungi for food have also multiplied through cheesemaking, leavening bread, brewing beer, and making wine. By domesticating the fungi that enrich our diet, we have folded the natural environmental actions of these microbes into human culture and driven changes in civilization over thousands of years. All of the uses of the microscopic yeasts and molds for food are extensions of the human-fungus symbiosis.

Human interactions with fungi reach further through the biotechnological manipulation of fungi in drug manufacture. Fungal medicines include antibiotics to treat bacterial infections; cyclosporin, which prevents the rejection of transplanted organs; and "human" insulin and vaccines produced by genetically modified yeast. On the subject of fungal medicines, there is a lucrative global market for medicinal mushrooms with reputed life-enhancing and lifesaving benefits. Consumers spend tens of billions of dollars per year on mushroom extracts in the belief that they are effective at treating illnesses through their effects on the immune system. Few of the claims made by marketers have been tested, yet there is some hopeful news on this front with investments in clinical trials on mushroom products as anticancer agents. Drawing on stronger scientific evidence, "magic" mushrooms have outgrown their countercultural associations to be embraced as a promising treatment for clinical depression and post-traumatic stress. Through these ancient and modern cultural practices, we have amplified the positive influence of the fungi on our lives.

FROM WOMB TO TOMB: THE FUNGUS IS WITH YOU

Some years ago, when I gave seminars on the so-called toxic molds found in buildings, I would say, "We inhale their spores from first breath to last gasp." This was true, but more recent findings show that the phrase provides an incomplete picture of our interactions with fungi, which begin before birth and extend into the grave. The bookends must be extended because the fungi are with us from womb to tomb.

The view of the womb as a perfectly sterile incubator for the fetus has been consigned to medical history today.[14] Our interaction with bacteria and fungi before birth is evident from genetic analysis of the meconium, which is the tarry liquid expelled by the newborn that develops in the fetal bowel. It is the initial deposit that comes to light in the first few days after birth and presages a lifetime per capita unloading of twelve tons of feces. Fungi that live within the mother's vagina, particularly the *Candida* yeasts, are the commonest species detected in the meconium, which suggests that they colonize the amniotic fluid that surrounds the fetus and make their way into the gut of the baby when it swallows a little of the fluid before birth.[15] Whether the traces of fungi found in the meconium affect the developing fetus is unclear, although we do know that the risk of premature birth increases as the number of microbes within the amniotic sac rises.[16] Fungi may also spread all the way from the mother's skin to the fetus through her bloodstream. Microbes are normally excluded from the blood, but when the tissue barriers are weakened, and leakage occurs, the bloodstream becomes a distribution system for bacteria and fungi. Inflammation of the gums or gingivitis, which develops into more serious forms of gum disease, or periodontitis, is one way that this can happen, raising the importance of oral hygiene and dental care for pregnant mothers.

Birth propels the baby into the microbial world, where it will be coated with fungi, filled with bacteria, infected by viruses, and breathe air brimming with all manner of irritating particles. As a fetus we receive sips of foreign proteins through the placenta and encounter the tiny populations of microbes in the amniotic fluid. Outside the womb we are deluged with microbes. Natural selection shaped Earth as a jungle of hospitable and hostile microbes over billions of years and each of us is forced to adjust to this carnival. The only possibility for survival resides with our guardian angel—the immune system—which allows us to explore this melee safely, cultivating microbes on the body that keep us healthy and warding off the germs that are more likely to kill us.

The deeper relationship with the microscopic starts after birth and depends on how we are born. If we surface via the natural route, the rupture of the amniotic sac leaves us squeezing through the vagina, where we receive a coating of bacteria and fungi from our mothers. The elastic fit of the birth canal around the baby ensures that the whole surface of the newborn receives this treatment. This is why the same strains of *Candida* yeast that grow in the mother's vagina are found on the skin of infants born in this manner. Babies born by C-section become colonized by *Candida* from their mothers too, but by different strains of the yeasts that are more abundant on the surface of the mother's skin. Difficulties in identifying fungal species make these studies challenging, but the overall pattern of vaginal microbes for vaginal deliveries and skin microbes for the C-section babies seems to be true. Our birthday is the first day of intensive schooling for the immune system that has been preparing for a microbiological hurricane for nine months.

Fungi are also conveyed through intimate contact with the mother and handling by others after delivery, and breastfeeding furnishes the infant with a different mix of species. According to recent experiments that took precautions to avoid contamination from the nipple surface, there are a lot of fungi in breastmilk. These studies show that a slurping infant absorbs more than two hundred million fungal cells a day, along with a similar number of bacteria.[17] The majority of the fungi are yeasts, although there are traces of molds whose spores drift in from the air. It seems likely that fungi found in the milk get there through the bloodstream, like the ones that cross the placenta. Cells in the immune system may also pick up fungi from the nasal passages, lungs, and gut and carry them to the breast tissue where they are released into the milk. The couriers that perform this proposed role are the dendritic cells that prowl the body looking for microorganisms.

The transmission of microorganisms from mothers to their offspring is a widespread phenomenon in nature.[18] The ubiquity of this seemingly deliberate donation of foreign organisms from mother to child suggests that some of them are critical for the vitality of the newborns of all animals. Most of these microbes appear to be beneficial, although some pathogens can also be relayed in this manner. Harmful microbes that pass from the human mother to child include the parasite that causes toxoplasmosis, the syphilis bacterium, and the human immunodeficiency virus (HIV). The fact that some pathogens take advantage of this intimacy is an inevitable drawback to the elementary processes that support the next generation. Natural selection has fostered microbe transfer because the advantages of readying the immune system for the outside world outweigh the disadvantages. The proof is evident in the success of the human reproductive mechanism and the fact that most infants do not succumb to microbial infection. Evolution is blind to a small proportion of casualties.

The split personality of the fungi as beauty and beast is exactly what we would predict from nature. The idea of some grand harmony among organisms is pure fantasy. Like every other scrap of life, fungi are engaged in a continuous struggle for existence. They collaborate with other organisms when there is mutual benefit, fight when their partners act too aggressively, and set up shop elsewhere by dispersing spores. Others have no time for collaboration and live by damaging host tissues from the moment they arrive. When we think about the beneficial fungi on the body and inside the gut, there is a tendency to ascribe intention to the relationship. This is wishful thinking. The fungi that grow on us fulfill a role that they have crafted, occupying a niche on the body—the scalp for example—where no other microbes do it better. During their evolutionary history, these fungi have developed the mechanisms for living on the dryness of the head, feeding on fats, and keeping the bacteria in check. As long as they do not irritate the skin, they are left in peace. When the environment between the hairs becomes unbalanced in some way, the fungi get unruly, and accumulate in unusual numbers. This causes the skin to flake, and the immune system is alerted that something is amiss. Swings and roundabouts, day in and day out, ups and downs, and the same goes for the digestive system, vagina, and everywhere else that the fungi blossom on the body.

Fungi that adapt themselves to living in these different locations in the human ecosystem are the ones that need to be recognized as partners by the immune system and left alone. Much about this process of acceptance remains unknown, but the immune system of the baby seems to be tutored inside the uterus, during birth, and via breastfeeding, priming the newborn for a lifetime of encounters with microbes. Learning to work with some of the commonest kinds of fungi on the skin seems like a very good lesson. By acknowledging these harmless species and avoiding conflict, their growth may limit the ability of bacteria to coat the skin. We know that yeasts engage in chemical warfare with bacteria, and as long as these fungi do not proliferate to the point of becoming a nuisance themselves, they remain part of the healthy microbiome.

The newborn immune system is very fragile, with a small population of mobile cells to police the lungs and other tissues, but the number of these caretaker cells skyrockets to adult levels within the first days of life. This period of vulnerability to infection is one of the characteristics of infants that resulted in such a high level of historical mortality. Before the twentieth century, more than 25 percent of babies died in the first year of life, and half of all children died before they reached the end of puberty. Oral *Candida* infections, or thrush, which spread to other parts of the body were a significant cause of infant deaths.[19] A combination of better nutrition and hygiene, and the development of vaccines, antibiotics, and other drugs in the twentieth century transformed this picture, and today's infant mortality rates range from 5 percent in some African countries to less than 1 percent in Europe.

Considering these statistics, it is difficult to find fault with modern medical practices, but we have introduced new hazards by increasing the number of births by C-section and avoiding breastfeeding.[20] Breastfeeding follows a disturbing pattern, with few countries meeting the WHO and UNICEF targets for infants.[21] These regional differences are affected by ethnicity and culture as much as economic prosperity. They are alarming from a mycological perspective because the

use of milk formula is linked to a greater risk of asthma and other allergies to fungi. The immune system needs to be introduced to grubbiness
of the natural world from the get-go, and if this lesson fails, we may
face a lifetime of struggle, responding to each fleck of fungus as if we
are under attack. There is some evidence that this poor formatting of
the immune system in childhood can also lead to autoimmune diseases
later in life.[22]

The drama of the mycobiome does not end with infancy, and our
interactions with fungi change as the years pass. The mycobiome in the
digestive system is adjusted in response to the shift to solid food and
later as we pursue more omnivorous, carnivorous, or vegetarian diets.
The mycobiome is modified in adulthood, wobbling this way and that
as we gain and lose weight, become pregnant, undergo cosmetic surgery
and dental procedures, take antibiotics, are injured, and develop short-
term (acute) or long-term (chronic) illnesses. We can pick up fungal
infections at any time, although the likelihood of developing these mycoses increases among the elderly. Fungal infections increase with age
as we develop illnesses that damage the immune system or require us
to undergo therapies that lower these defenses to maintain transplanted
organs or to treat cancer.

Moving to the tomb, we are the latest additions to the menu of meat
dishes available to fungi that have been in the recycling business for
hundreds of millions of years. Bacteria take care of a lot of the soft
tissues in the human cadaver, leaving the keratin protein in hair and
nails for a variety of specialized molds.[23] Bones are infiltrated by fungal hyphae too, assisting their disintegration. These are the slowest
processes of decomposition, but the disappearance of the hair, nails,
hooves, horns, and antlers of dead animals in nature demonstrates the
power of the fungi. We are sustained by this planetary-scale nutrient
cycling, together with the breakdown of plant debris, because we rely
on the forests, grasslands, and agricultural ecosystems fertilized by the
fungi. We would not be here without them. Mycorrhizal partnerships
with living plants are another essential part of this broader relationship.
We need fungi to work with plants as much as we need fungi to work
with our bodies.

THE PAGES AHEAD

In the chapters that follow, we will explore the communities of fungi that live in every place on the body, pursue the historical and contemporary uses of fungi as food and medicine, and look at the ecological roles of the fungi beyond the body that function as an invisible life-support system. This book is a revelation of the human relationship with the fungi, the human-fungus symbiosis, from the fungi growing around the roots of our hairs to the colonies of mushrooms wrapped around the roots of forest trees. The chapters are organized into two sections that speak to the direction of the inquiry: part I, "Inward," and part II, "Outward."

Part I, Inward. Starting with the fungi on the skin surface (chapter 2), we slip inside the body with the spores that enter the lungs (chapter 3), and dive deeper with the yeasts and filaments that infect our internal organs (chapter 4). All tissues in the body can become colonized by fungi, which expands the activities of the restless mycobiome to the lungs, liver, kidneys, brain, and gut. Fungi in the gut are also part of the healthy digestive system, until this part of the mycobiome is disrupted and a whole range of illnesses related to immune dysfunction develop (chapter 5).

Part II, Outward. In the second set of chapters, we look at our interactions with fungi that grow outside the body, beginning with their importance in our diet, both as wild and cultivated mushrooms, and in the development of mycoprotein meat substitutes that are energizing the food industry (chapter 6). This concept of the extended symbiosis incorporates the modern use of drugs produced by genetically modified fungi in conventional medicine and the controversial marketing of mushroom extracts in alternative or naturopathic medicine (chapter 7). From medicines we move to poisons and the dangers of consuming the wrong mushrooms and the spoilage of grains by molds that produce mycotoxins (chapter 8). Magic mushrooms and their use in the treatment of clinical depression and other serious mental health issues follows (chapter 9). Proposed links between the use of magic mushrooms and the origins of Christianity and other religions have been dismissed by theologists, but there is room to reconcile these viewpoints with a fresh and objective analysis of the evidence. In the final chapter we examine the ultimate extension of the human relationship with the fungi through our dependence on their wider ecological roles (chapter 10). We discover the body as an ecosystem flavored with fungi—a pulsing city or *mycopolis*—dependent on the fungi that support plants, create soil, filter rainwater, and spin the carbon cycle.

This is the story of people and fungi, the human-fungus symbiosis that spans the local to the global, revealing how our lives are influenced through rich relationships with these extraordinary microorganisms.

PART I

Inward

2

Touching

FUNGI ON THE SKIN

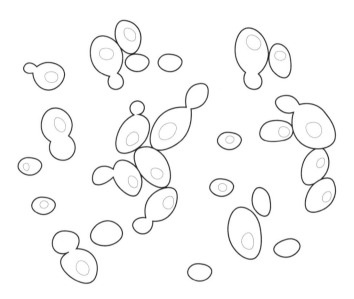

ALTHOUGH WE MAY be exposed to fungi in the womb, the coating of yeasts that forms when we are born marks the real beginning of the life-long human-fungus symbiosis. Fungi enter the lungs and the digestive system as soon as we start breathing and breast- or bottle-feeding, but the skin remains the biggest territory for the fungi throughout life, the place where they dominate our microbiology. The skin is considered the largest human organ, and the fungi grow all over it, consuming natural oils and dead cells, supporting and irritating the folds and furrows of the external tissue or epithelium. They grow in the greatest numbers on the scalp, where one hundred thousand to one million yeasts can huddle in the space of a postage stamp.[1] If humans were squeezed together at this density, all eight billion of us would fit into a city the size of Los Angeles.[2] When we look in the mirror and brush our hair, we have no sense of this congestion, but the fungi are in full swing, stirring the chemistry of the skin, bossing the tinier bacterial residents around, and causing the tissues to redden and flake when their routines are disturbed by a new soap, shampoo, or lotion.

Most of our modesty can be concealed with a bath towel, but the skin surface available for microbial growth is more extensive, matching the area of thirty towels if we perform the thought experiment of flattening out the nooks and crannies of the five million hair follicles.[3] The populations of fungi on this landscape are adjusted from birth to death, with yeasts and filamentous species coming and going according to rules that we are only beginning to understand. The numbers and kinds of fungi that grow from head to toes have also changed throughout history as we have wrapped ourselves with clothes, slipped on shoes, and doctored the environment with cosmetics and drug treatments. Going back even farther, the skin mycobiome has been making and remaking itself since modern humans emerged from the Rift Valley of Africa.

The skin is not the most inviting place for microorganisms, because food and water can be scarce. These challenges have led fungi that live on the scalp to specialize in feeding on the waxy sebum secreted from sebaceous glands and others to become very good at breaking down the keratin protein in the outermost layers of the skin. Perspiration provides salty water, and some of the residents overcome the aridity by producing their own water as they digest the fats in the sebum. Through these measures, yeasts and molds luxuriate on the skin. There are even more bacteria on the surface of the body, but this is where the dimensions of the fungal cells become pivotal. Although there are ten bacteria for every fungus on the skin, the fungi outweigh the bacteria by a factor of ten.[4] This difference in size explains why the fungi are so important to the ecology of the skin. Fungi also abound in the gut, as we will see in chapter 5, but they do not fare quite as well as the bacteria. One of the reasons for this is that fungi like to be flushed with oxygen, which is quite limited in the intestines. Many of the gut bacteria are more flexible in their oxygen requirements, which explains their growth in the trillions.[5]

Understanding what fungi do on the skin surface is a work in progress for experts on the microbiology of the skin, with more questions than answers, and a lot of conflicting information about the identity of the fungi that support the clearest complexions, most luxurious hair, and healthiest nails. The importance of quite subtle changes in the skin mycobiome is illustrated by a complaint known as sensitive skin syndrome. This skin condition is very common, affecting more than half of all people, if we include very mild cases. Symptoms are subjective, making it difficult to diagnose, and include stinging, burning, and tingling sensations that follow the use of cosmetics and exposure to everyday irritants. There are no visible signs of sensitive skin in most patients, but when reddened patches appear we call this erythema. The mycobiome was implicated in sensitive skin in a study from South Korea that found a greater diversity of fungi in skin swabs taken from women with the syndrome relative to control subjects.[6] *Malassezia* yeasts were the most frequent fungi swabbed from the cheeks of all the women, but in the patients with sensitive skin, this yeast was diluted by a surge in the growth of other kinds of fungi, including a mold called *Mucor*. The mycobiome differed from case to case, with little uniformity between the communities of fungi that developed. Fungal involvement in the chronic skin inflammation in sufferers of psoriasis follows the same pattern as sensitive skin syndrome, with a greater diversity of fungal species found in the patches of damaged skin compared with the intervening areas of healthy skin.[7]

This research shows that these skin conditions are associated with disruptions to the normal mycobiome. Dysbiosis is the term used to describe instances of microbial disturbance, whether they are associated with disease or not. Turbulent mycobiomes and microbiomes more generally are part of the normal pandemonium of nature, which makes it doubly difficult to determine when the appearance of an unusual fungus means that something is amiss. Although skin inflammation can be a direct response to the growth of a particular fungal species, we do not refer to complaints like sensitive skin syndrome as infections. Diagnosis of a fungal infection, or mycosis, requires a greater level of tissue damage, but we are dealing with a continuum of symptoms associated with fungi on the skin rather than a clear distinction between an unsettled mycobiome and more problematic disease.

At both ends of the spectrum of fungal development on the skin, the behavior of the mycobiome is affected by the response of the immune system. The immune system has a definite role in shaping the mycobiome, by permitting some fungi to grow and eradicating others. For its part, the mycobiome trains the immune system to recognize harmless and harmful adjustments in numbers and species. When we look at the most serious mycoses, we often find that they develop after damage to the immune defenses. Infections of our internal organs are featured in later chapters, but the mycoses of the skin arise from our continuous interactions with fungi in the environment and happen to people with perfectly healthy immune systems.

RINGWORM, ROBERT REMAK, AND RADIATION

Greek and Roman physicians were familiar with ringworm, which they lumped together with other scalp conditions and called *porrigo*.[8] Public bathing in ancient Greece and Rome and the practice of pouring olive oil onto the skin and removing it with a curved blade or strigil was a surefire remedy against skin parasites.[9] But the oiled skin was an invitation for the growth of fungi, and ringworm was a common complaint in Rome. The first emperor, Caesar Augustus, was blemished with "a number of hard dry patches resembling ringworm, caused by vigorous use of the scraper on an itching skin," and the affliction may have factored in the suicide of Roman senator Festus, who was desperately ashamed by "the deformity of a Ringworme in his face."[10]

Ringworm is a general term for fungal infections that can take hold on any area of the skin. Early references to "rynge-worme" appear in the 1400s, and toward the end of the seventeenth century John Aubrey, the writer and pioneering archaeologist, noted the resemblance of the skin lesions to the fairy rings produced by mushrooms in his *Natural History of Wiltshire*: "As to the green circles on the downes, vulgarly called faiery circles (dances), I presume they are generated from the breathing out of a fertile subterraneous vapour. (The ring-worme on a man's flesh is circular. Excogitate a paralolisme between the cordial heat and ye subterranean heat, to elucidate this phenomenon.)"[11] In other words, if we believe that fairy rings arise from the release of poisonous fumes from the earth, we might get at the root cause of ringworm by examining the humors of blood (air), phlegm (water), black bile (earth), and yellow bile (fire), on which the medicine of Aubrey's day depended. This kind of deductive logic is the reason that Aristotle is such a poor guide to the workings of nature, although a few classical scholars may disagree with this brusque assessment. In any case, while Aubrey's observations did not lead to any treatments for ringworm, he was spot-on about the similarities between fairy rings and the skin disease. In both cases, the hyphae of filamentous fungi start growing from single spores, striking outward in all directions at once to produce circular colonies.

More intentional investigations on disease-causing or pathogenic fungi began in the nineteenth century. Richard Owen, the famous anatomist, stirred interest in fungal infections when he discovered "a green vegetable mould or *mucor*" in the lungs of a flamingo during the dissection of a bird that had died in the London Zoo.[12] Owen concluded that the fungus was a parasite that had been growing in the bird before its death. His study was one of a scattering of early descriptions of the mycosis called aspergillosis, which also occurs in humans, but the first proof that a fungus was responsible for human disease came from research on ringworm. In 1842, twenty-seven-year-old medical researcher Robert Remak conducted a remarkable and unpleasant experiment by lifting a scab from the scalp of a patient suffering from ringworm and taping it to his own forearm.[13] After two weeks, he noticed "a strong itching . . . [and] found a dark red spot the size of a vest button" at the inoculation site. Removing the crusty spot, he found the fungus embedded in his skin, proving that it was responsible for ringworm.

Remak's exploration of ringworm belongs to a tradition of professional commitment by medical researchers that includes deliberate self-infection with the bacteria and viruses that cause stomach ulcers, yellow fever, relapsing fever, and venereal diseases, as well as the consumption of radioactive dyes and self-catheterization of the heart.[14] Some experiments in this vein led to Nobel Prizes, but the ethics of self-experimentation are problematic, may contravene the Hippocratic oath, and are assuredly at odds with modern guidelines for good clinical practices. Although his experimental approach was questionable, Remak clearly established that a fungus could cause an infection, and, remarkably, he had made this breakthrough twenty years before Louis Pasteur linked microbes to disease in his germ theory. Remak cared very little about scientific honors—the glittering prizes revered by most members of the academy. He was an unusually humble scientist who refused to take credit for his ringworm discovery, insisting, instead, that an older colleague had made the crucial insights. Today, he is recognized as one of the pioneers of medical mycology, which is the study of fungal infections and their treatment.

Fungi that cause ringworm feed in the outermost layers of the skin, where older cells are pushed to the surface by the underlying tissues. These cells are stuffed with fibrous keratin protein and become embedded in a fatty matrix as they die. This structure has been likened to a brick-and-mortar wall. It forms a shield against dehydration from the inside and infection from the outside and is constantly renewed as the oldest cells are shed into the environment. Hyphae of the ringworm fungi invade this skin layer, releasing enzymes, digesting the protein and the fat, and expanding into their fairy rings. Hair and nails are also fashioned from keratin and can become infected with ringworm fungi. In scalp ringworm, the fungi dive into the hair follicles and invade the hair shafts, feeding on the keratin until the hairs become brittle, fracture, and fall out, leaving bare spots on the scalp.

Ringworm fungi belong to the ascomycete group, which includes species of *Penicillium* that produce antibiotics, and *Aspergillus* that cause the lung infection found by Owen when he dissected the flamingo.[15] Ringworm infections are dubbed with Latin names according to the sites where they grow and other distinguishing characteristics—*tinea capitis* for the scalp, *tinea pedis* for the foot, *tinea unguium* for the toenails, and so on. Other skin infections come under the umbrella of *tinea corporis*, which includes a mycosis that spreads between young wrestlers and judo students that has been given the splendid name *tinea corporis gladiatorum*.[16] *Tinea corporis gladiatorum*! If dermatologists pursued more of this kind of creative nomenclature, their patients might feel a modest elevation upon their diagnoses: "athlete's foot" is a bit deflating, so, how about *tinea pedis-athletarum, -gymnasticorum,* or *-victorum*? Just an idea.

Ringworm infections are the commonest type of mycoses, affecting one billion or more of us at any time.[17] Ringworm is an affliction of childhood, and the number of cases falls abruptly after puberty. This suggests that the fungi are rebuffed by the hormonal hurricane that alters the chemistry of the skin secretions, reconditions the immune system, and leaves most of us feeling quite unsettled for a while. Ringworm infections remain prevalent in sub-Saharan Africa, where a Kenyan study found that 81 percent of children from an "informal settlement," or slum, in Nairobi suffered from *tinea capitis*—the scalp infection.[18] A separate survey in rural Nigeria showed that almost half of the children in the ethnic Nok community were infected.[19] The common name for *tinea capitis* used by the Nok is translated as "spider web," which refers to the belief that spiders urinate on the heads of children, lay their eggs, and create the ringworm patterns of concentric rings when they spin their webs. At the time of the study, published in 2016, the local barber was shaving the heads of the children with the worst cases of ringworm. He sterilized his clippers between shearings by dousing the blades with denatured alcohol and igniting them with a cigarette lighter.

Poverty and poor hygiene have always been a magnet for ringworm infections, and they were very widespread in Europe when Robert Remak showed that they were caused by fungi. Compared with the early treatments recommended by Western experts, the Nigerian barber's remedy seems very gentle. Complete hair removal was a common treatment, which involved plastering the scalp with molten tar or resin. This paste was left to harden into a solid cap that was ripped away, carrying infected and uninfected hairs as well as skin tissue and, I imagine, leaving the poor child blinking in horror and traumatized for life.[20] This seems to have been a popular remedy in the nineteenth century. Doses of thallium acetate, or rat poison, produced the same effect, although the drawback of prescribing an oral medication whose "therapeutic dose is so near the limit of the lethal dose" could not be ignored when children started dying.[21]

Ringworm spreads from person to person via infected skin cells, and outbreaks became common during the industrial revolution among children living in the "rookeries" or slums of London and other over-crowded cities. Orphanages and boarding schools—think *Nicholas Nickleby*—were breeding grounds for fungi, and then, in the 1890s, there was a medical breakthrough in the treatment of ringworm: X-ray epilation. Hair loss was observed in people treated with X-rays for other skin conditions, so the application of the technology for deliberate hair removal seemed an obvious step. The effectiveness of the method was undeniable. What could possibly go wrong? *The Lancet* published a let-ter in 1896 suggesting that gentlemen should zap their chin hairs with X-rays for a few minutes every evening to save the effort of shaving the next morning.[22] This practice failed to catch on, but women visited "Tricho" salons where cosmetic X-ray machines were used for removing unwanted hair until 1929.

The danger of the procedure was evident from the fact that the inven-tor of these appliances went on to have his left hand amputated to stop the spread of cancer and experienced the loss of his right hand to ulcer-ation.[23] But enthusiasm for curing ringworm with radiation continued in the 1930s, when a distinguished London radiologist wrote, "Even if, through some error in technique, an overdose is given, the worst that can happen is an X-ray burn."[24] This reckless attitude persisted for decades, and hundreds of thousands of children were treated before the therapy fizzled out in the 1960s. Radiation exposures varied, and the redirection of the beams toward different parts of the head during a treatment session reduced the damage to any single spot. Nevertheless, there is no doubt that children received X-ray dosages that we reserve for the treatment of brain tumors today, and which, tragically, may have caused brain tumors in an untold number of these patients in later life.

The frequent repetition of the word "may" in this book (and other expressions of possibility rather than certainty) is demanded by the sci-ence. In the ringworm X-ray story, we are not sure that the treatment resulted in the development of cancer in adulthood. It is very difficult to tie individual cases of cancer to a particular exposure to radiation, and so we pursue epidemiological studies in which we gather information on as many ringworm patients as possible and see what happened to them in later life. If there is a clear spike in brain tumors in these patients compared with untreated controls, we edge closer to scientific certainty. If the number of cancer cases is no greater, or not much greater, than the background, we are left with an unresolved problem. Some of the studies on ringworm patients suggest that there is a link between the use of radiation and cancer, while others do not.[25] It is possible that we will never know for sure. Besides the fear of a cancer diagnosis, some Israeli girls treated with X-rays suffered from permanent hair loss that left them with lifelong cosmetic and psychiatric challenges.[26] However we look at it, this ringworm treatment was a low point in medical mycology.

Most ringworm fungi provoke a relatively mild response from the immune system, and some children with disfiguring skin infections are spared from intense itching and other symptoms of inflammation. This blessing is a feature of the limitation of fungal growth to the layers of dead skin cells and separation from the deeper tissues patrolled by the active cells of the immune system. The ringworm fungi also use various tricks to camouflage themselves while they feed on the skin. One of these mechanisms allows the fungus to conceal molecules on the surface of its hyphal filaments that otherwise serve as alarm bells for the immune defenses. This invisibility cloak evolved over tens of millions of years, as soil fungi adapted themselves for growing on the skin of different kinds of animals. Relatives of the human ringworm fungi colonize other mammals without causing any obvious harm, but they can damage our skin if they are transferred from farm animals or pets. Infections that come from other animals are called zoonoses, and they are a good reason for avoiding contact with hedgehogs. Intimacy with hedgehogs does not concern most of us, but cases of ringworm infections among lovers of these spiny animals are surprisingly common. Dogs and cats are more common sources of human ringworm, and the ubiquity of these pets makes these zoonotic infections inevitable.[27] Islamic prohibitions against keeping dogs in the home seem entirely reasonable from a mycological viewpoint.

Fortunately, the development of antifungal medicines means that ringworm infections can be cured today without recourse to violent epilation, rat poison, or X-rays. Griseofulvin was discovered in the 1930s, and its use for treating ringworm began in the 1950s. It is made naturally by a species of *Penicillium* that causes blue mold on harvested apples and is a close relative of the fungi that produce the penicillin antibiotics. Griseofulvin is taken as an oral medication and kills the ringworm fungi from the inside out, making its way to the skin surface where it infiltrates the hair follicles and destroys the hyphae from the bottom of the hair shafts toward the tips. It works by disrupting the division of nuclei inside the hyphae. Other antifungal drugs effective against ringworm include terbinafine and azole antifungals that disrupt fungal membranes. Apple cider vinegar, tea tree oil, and raw honey are some of the natural products promoted as alternative treatments for ringworm. (We look at antifungal drugs in more detail in chapter 4.)

DANDRUFF

Outbreaks of scalp ringworm in children are uncommon in more prosperous countries today, and the drug treatments are effective in treating individual cases when they develop. This does not mean that the scalp has become a microbiological desert. Far from it. It is a hive of fungal activity throughout our lives, no matter how many times we wash our hair. Most of the fungi that grow on the skin do so as yeasts rather than molds, as blobs rather than webs. This is a good thing, because yeasts stay on surfaces, whereas molds, or filamentous fungi, are fashioned for penetrating tissues, and nothing good comes from skin invasion by fungi. Species of *Malassezia* yeasts are the dominant fungi on the scalp. They are named for a French anatomist, Louis-Charles Malassez, who found them growing in skin flakes scraped from patients suffering from seborrheic dermatitis. Seborrheic dermatitis is an extreme form of dandruff, sharing many of its characteristics with the snowiness of hair that afflicts a good chunk of humanity to varying degrees. Both complaints involve the multiplication of *Malassezia* in the sebum exuded from the sebaceous glands. The mouths of these microscopic glands open into the hair follicles wherever we are hairy, and directly on the skin surface in places where we are not. Sweat glands are separate things that release more watery secretions whose evaporation is key to controlling body temperature.

Sebum is marvelously complicated stuff that contains a mélange of fats and oils and is produced in varying amounts according to age and sex—more in men than women—and serves as the dietary staple for the yeasts that live on the skin. These fungi are so perfectly adapted to life on the skin that they have lost the ability to produce their own fatty acids like other organisms and draw everything they need from the sebum. We consume fats, of course, but our ability to manufacture fatty acids from sugars in the diet is essential for constructing membranes and performing all kinds of other metabolic tasks. By surrendering this almost universal biochemical capability, the evolving yeast saved a great deal of energy and bonded itself to the skin for the rest of forever.[28] *Malassezia* belongs to the basidiomycete group of fungi rather than the ascomycetes that include the molds that cause ringworm. Fungi that form gilled mushrooms are classified as basidiomycetes, but the closest relative of the dandruff yeast is a fungus that causes a crop disease called corn or maize smut. (The infected ears become filled with blackened spores that are used as an ingredient in Mexican cooking called huitlacoche.) Both of these fungi—dandruff yeast and corn smut—are specialized organisms that have become completely dependent on their hosts.

Dandruff is an inflammatory condition that develops as the yeast works its way into the skin, feeding on the sebum and releasing irritating compounds onto the scalp. This disturbance to the skin chemistry alerts the immune system, which responds by mobilizing macrophages and killer cells against the fungus. Itching and skin flaking are symptoms of the unfolding turmoil on the scalp. *Malassezia* lives on everyone, so the reason that some of us are spared dandruff and others itch, scratch, and flake is a bit of a mystery. What we do know, however, is how to treat it.

Early in my research career, I worked at Yale University with a visiting scientist from the Soviet Bloc who was very careful with money, saving as much as he could from his salary to keep him in relative comfort when he went home. To this end, he collected sachets of ketchup and mayonnaise from fast-food restaurants rather than purchasing these condiments from the grocery store. Dandruff was a significant problem for this expert on fungal physiology, and rather than wasting money on the medicated shampoo that I recommended, he set off to find some stinging nettles, which, he explained, are a natural balm for all scalp problems. Finding a patch of nettles behind our lab building, he boiled the leaves, mixed them with vegetable oil, and before long his hair shone like Samson's mane. An alternative remedy chosen by more than one billion dandruff sufferers comes in plastic squeeze bottles filled with the best-selling shampoo in the world, namely, Head & Shoulders, manufactured by the Procter & Gamble Company. This lucrative product has been on the market since the 1960s.

Dandruff shampoos kill the dandruff fungus with various formulations containing pyrithione zinc, selenium sulfide, and piroctone olamine. I am detailing the names of these chemicals so that you can look at the small print on your shampoo bottles and see which ones you are lathering into your hair. They kill the fungus by messing up its membranes, which either starves or poisons the cell.[29] The control of dandruff is a triumph of Western science. Not as spectacular as antibiotics or vaccines, but something to smile about when you grab the shampoo in the drugstore. I have begun to wonder, however, if there may be a downside to this pharmacological battle against the yeasts that have lived peacefully on the human scalp for millennia.

If we use the guesstimate of two hundred thousand years for the origin of our species, the natural symbiosis between humans and the scalp yeasts endured for 99.97 percent of our partnership before we began killing them with shampoo. If a single species of *Malassezia* was the sole occupant of the skin microbiome and we struck it down with shampoo in the pursuit of lustrous unflaked hair, there would be little else to say. But the scalp is a more complex ecosystem, where multiple kinds of yeasts are found, filamentous fungi show up with some regularity, and both kinds of fungi share the neighborhood with bacteria.[30] These microorganisms work with one another, and against each other, rising and falling in numbers as the bearer of the scalp moves from childhood to adolescence and onward to adulthood and old age. The daily use of antidandruff shampoos does not seem to cause any side effects, and we certainly feel blessed by the absence of itching and flaking if we have experienced the alternative. But another fungus that grows on the skin has made me think a little deeper about the consequences of manipulating the mycobiome.

AN EMERGING AND DANGEROUS YEAST

In 2009, a new kind of yeast was discovered in the ear canal of a seventy-year-old woman in a Tokyo hospital. The DNA signature of this fungus was sufficiently different from other species that it was given a new name: *Candida auris*. Since then, this yeast has become a global plague—plague may be a term too fearsome, although it is infecting and killing hospital patients across the world and is resistant to every type of antifungal medicine. *Candida auris* grows on the skin of patients who pick up the fungus in hospitals where it has already made itself at home on other patients and sticks to the surfaces of medical devices.[31] As long as it remains on the skin it does not do any harm. The problems unfold when the yeast makes its way into the bloodstream through catheters inserted into veins to deliver fluids or drugs. Once inside the body, it multiplies by shedding buds from its cell surface and spreads to the kidneys, heart, brain, and other organs, causing fever, breathing problems, and fluid retention. The mortality rate for the worst infections approaches 60 percent. In the same year that this new species was described in Japan, reports of hospital infections caused by the yeast came from South Africa and India, and then, in the following year, in Kenya. Soon, the fungus was spreading across dozens of countries.[32]

Genetic analysis has shown that there are four distinctive populations of *Candida auris* and that this fungus evolved very recently in biological terms.[33] The estimated timeline for this evolutionary process comes from changes in the DNA sequences of yeasts belonging to the different populations. The kinds of genetic variations identified in these studies develop at a relatively constant rate, and so the number of alterations in the DNA sequences serve as a molecular clock for researchers. This clock suggests that the oldest population of *Candida auris* emerged in the middle of the seventeenth century—around the time of the Great Plague and Great Fire of London, and that the strains of the fungus that cause the most aggressive infections developed as recently as the 1980s. Using the incredible power of whole genome sequencing, we can follow the pulse of the DNA of this pathogen across four centuries and witness its appearance as a major threat to human health. This is a remarkable piece of work. A second discovery shows that the yeast is not completely dependent on human skin, because it has been found in sediment samples from a salt marsh and a sandy beach on the Andaman Islands in India.[34] How and why it began killing humans is not known, but there are some clues.

One provocative suggestion is that global warming has led to the evolution of hyperaggressive yeast strains that can thrive in the warmth of the human body.[35] This seems unlikely, because the increase in average global temperatures has been too small to push the fungi beyond their existing comfort zones. The majority of fungi, which we call mesophiles, grow fastest at 25°C–30°C (77°F–86°F), but many of them can keep growing until the temperature approaches 40°C (104°F). Our body temperature is challenging for these fungi rather than crushing. On the other hand, shifting weather patterns related to global warming may have a significant effect on the distribution of hotspots for fungal asthma (see chapter 3) and outbreaks of life-threatening mycoses (see chapter 4).

Setting aside its temperature tolerance, other alterations to the ecological experience of *Candida auris* may be more significant in its recent development as a pathogen. We know that the use of antifungal drugs introduced in the 1980s has driven the emergence of resistant strains of other species of *Candida*, and this process might apply to *Candida auris*.[36] These medicines, which include ketoconazole and fluconazole, are examples of the azole antifungals that continue to be used to treat all kinds of fungal infections. Azoles and other antifungal agents are also used to control fungal diseases of crop plants, and their presence in the environment is likely to drive the emergence of resistant strains of fungi that could attack our food supply and our bodies.[37]

The use of dandruff shampoo is so pervasive that we do not think of it as an antifungal agent, but many of us use this as a daily mycobiome disrupter. *Malassezia* is the unchallenged monarch of the skin until it is drowned with toxins when we stand in the shower and lather our hair. The antifungal agents in these hair products cause monumental changes in the ecology of the skin, and a fungus that can rise to the challenge by surviving the shampoo will prosper in the vacuum left by Queen *Malassezia*. This is natural selection pure and simple. Precisely the same thing happens when an antibiotic is used to treat bacterial infections: a strain of the bacteria that develops resistance takes over from its defenseless relatives, offering the classic example of evolution by natural selection. This process explains the emergence of lethal bacterial infections that do not respond to antibiotic treatment. So, along with the proliferation of antifungal medicines and agricultural fungicides, the popularity of dandruff shampoo should be considered as a global invitation for the evolution of worrisome strains of fungi that can interact with the human body. We do not know why *Candida auris* became a problem in the 1980s, but the answer may lie somewhere in the growing use of antifungal products.

ATHLETE'S FOOT: THE DOWNSIDE OF SHOES

We have remained at the head end of the human-fungus symbiosis for most of this chapter, and now we move to the feet for the climax. To find the earliest case of fungal infection caused by an otherwise life-enhancing invention, we have to look deeply into human history and examine a 5,500-year-old shoe discovered in an Armenian cave.[38] The shoe was made from a single piece of animal hide, which was wrapped around the foot and laced together on top. This is the oldest known closed shoe, and the essence of the design is responsible for athlete's foot. The Armenian shoemaker was embarking on an unconscious mycological experiment with a wearable consumer product, much like the later inventors of contact lenses, which come with their own fungal complications.[39] While a closed leather shoe provides warmth in temperate climates, protects the wearer from sharp objects on the ground, and, let's face it, looks very stylish, it can create a hot and humid chamber that is as close to perfection for cultivating fungi as a throbbing stainless-steel incubator in a mycology lab. The result, for susceptible individuals, is ringworm of the foot, *tinea pedis*, which may be the most widespread fungal infection of our species.

The Romans we met earlier who developed spots of ringworm on the rest of their oiled bodies avoided the foot complaint by wearing open sandals. Sandals also protected Latin wearers from a dreadful fungal infection called mycetoma that is acquired by stepping on thorns bearing the fungus. Forced into the tissues, this terrible fungus bores interconnected channels or galleries throughout the foot that open onto the skin surface and discharge a watery fluid infiltrated with infectious grains.[40] Evidence of the disease has been found in the skeleton of a Roman who died in his late forties or early fifties in the second or third century AD. The bones of both of his feet have the moth-eaten look that is characteristic of this mycosis.[41] Mycetoma is a subtropical disease whose distribution suggests that he might have acquired the infection when he served in one of the North African provinces. Perhaps he took off his sandals to play the Roman ball game harpastum with the locals. The infection is also known as Madura foot, after the district of Madurai in South India, where cases caught the attention of colonial doctors in the nineteenth century. (Carrying bundles of wood on the head or shoulders leads to rare cases of Madura head. An image search for this awful disease should be avoided before bedtime.[42]) Sporotrichosis caused by the fungus *Sporothrix* is another infection that can begin with a thorn prick and is called rose handler's disease.[43]

Returning to the mycological perils of wearing closed shoes, the typical cause of athlete's foot is the fungus *Trichophyton rubrum*, which evolved in Africa. We know that this fungus arose in Africa because this is where the species harbors its greatest genetic diversity today.[44] This pattern of maximal genetic variation exists in the homelands of most species of all organisms for the simple reason that the process of mutation—which is the source of variation—occurs continuously and has more time to increase the diversity of a species where it originates than in locations where it arrives as smaller populations of migrants later in its history. The greatest genetic diversity in our species is found in Africa, the same continent as *Trichophyton*, and it seems certain that we left there together, fungus on foot, and spread across the planet. Ringworm may have been a minor problem before the exodus, growing on dead skin cells without moving any deeper. Closed shoes provided the opportunity for more luxuriant growth of the fungus, and some strains became more virulent as they encountered stronger resistance from the immune defenses. The same fungus is one of the species that infects toenails causing onychomycosis, although poor circulation in older patients is a major risk factor rather than simply wearing closed shoes.

Powders and creams for treating athlete's foot have a global market value approaching $1.7 billion.[45] This is not surprising, given the shelf space occupied by these products in pharmacies. Medicated shampoos are worth more than $12 billion, and dandruff is the most prevalent health complaint caused by fungi, although the fact that it is caused by a fungus is a detail that escapes most sufferers. The overgrowth of vaginal yeast, which we address in chapter 5, is a comparable problem in terms of the number of cases and far worse when we consider the symptoms. And, again, many sufferers may not think of yeasts as fungi. On the other hand, everyone knows that athlete's foot is caused by a fungus, which means that the space between our toes, the web space, is the most familiar interface between the human body and the mycological world. The discomfort caused by this everlasting pandemic is one of the most obvious expressions of our deep and evolving relationship with the fungi. More problematic, although less visible than reddened and peeling skin between the toes, is the inhalation of fungal spores that accompanies every breath. This is the subject of chapter 3.

3

Breathing

SPORES IN THE LUNGS

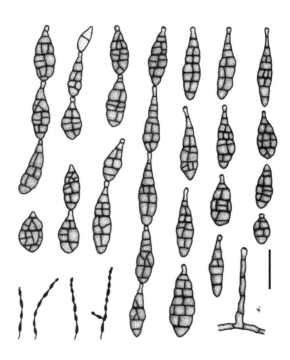

I AM OBSESSED with spores and have invested a sizable chunk of my professional life in understanding how they get into the air.[1] This admission is more likely to attract pity than interest until you consider the beauty of the dispersal mechanisms used by fungi. Start by lying on the grass next to a mushroom at night and using your phone flashlight to illuminate the underside of the fruit body. Move the light around until you see the smoke that pours and swirls from the cap. That

smoke is composed of hundreds of thousands of spores, which are propelled from the gills by little drops of water. There is grandeur in this
view of life.

Mushrooms are one of the sources of airborne particles that join the
clouds of spores released by the molds that grow on plants and every
other surface in nature. The air is filled with spores, and we inhale them
with every breath. These microscopic specks are destroyed after they
stick to the mucus in our airways, but they carry irritating proteins
called allergens that are as damaging to asthmatic lungs as birds sucked
into jet engines. Asthma and other illnesses that result from the inhalation of spores are the subject of this chapter.

Air seems to coagulate during an asthma attack, with each inhalation
urgently demanding conscious attention. Suffering from a severe bout
of asthma in England in July 1969, I spent hours bath-robed in front of
the television, watching the coverage of the Apollo 11 mission on the
BBC. As Armstrong and Aldrin explored the lunar surface, the pauses
in their conversations with Mission Control in Houston seemed to syn-
chronize with the laborious rhythm of my breaths, so that I began to
imagine that I was with them on the Moon. It was an oxygen-deprived
hallucination. Looking at the night sky after Armstrong's step and the
flag planting, it seemed that the Moon belonged to America. This evidence of the power of science convinced me that America was the place
to be, not this chilly island with its stifling air supply, but the land that
made all things seem possible.

Hamlet was struggling with ennui rather than breathing when he
described the air as "a foul and pestilent congregation of vapors," but
this is a perfect assessment of the atmosphere for an asthmatic. An
English physician, John Floyer, described the feeling of an asthma attack in his classic study of the illness, *A Treatise of the Asthma*, published in 1698: "The asthma is a laborious respiration, with lifting up of
the shoulders, and wheezing, from the compression, obstruction, and
coarctation [narrowing] of some branches of the bronchia, and some
lobes and bladders of the lungs."[2] Anyone who has experienced asthma
will recognize the "lifting up of the shoulders," which is an automatic
reaction to restricted airflow. Sit in a chair, keep your mouth closed,

pinch your nostrils until they start to flare on their own, and you will feel your shoulders rise after a few restricted breaths. Asthma reduces the space inside the lungs, and assuming a more upright position and raising the shoulders are subconscious strategies to force the airways open, to create more space. Doctor Floyer wrote from personal experience: "I have suffered," he wrote, "under the tyranny of the asthma at least thirty years."

Fungal spores are responsible for much of the tyranny of asthma. When we vacuum air though a filter and examine the harvest on its surface with a microscope, we find particles that look like tiny shards of broken glass, minute globes and ellipsoidal eggs, broken strands, fallen missiles, and juggler's clubs—a toy chest of the insane. These are the spores of fungi, made visible with the low power of the microscope, along with the larger pollen grains from plants. Bacterial rods and blobs appear at higher magnification, while the viruses remain invisible until we view the finest air filters with an electron microscope. We live in this soup.

Soup is an imperfect metaphor, because air is so thin and spores are so vanishingly small.[3] One hundred thousand spores per cubic meter is considered a very high concentration by experts on air quality; 10,000 spores is a moderate number, and 1,000 spores is very low. At rest, we take an average of twelve breaths per minute and inhale and exhale around six liters of air. This means that one spore is drawn into the lungs with every other breath when there are 1,000 spores suspended in each cubic meter of the surrounding air, five spores per breath at 10,000 spores per cubic meter, and so on. Some of the spores are immediately expelled when we exhale, others stick to the mucus in the lungs. Over a lifetime, this equates to lung contact with more than one billion spores, which is a lot of spores, but amounts to no more than the weight of a pea.[4] How on earth, one may ask, can so slight an interaction result in decades of suffering for an asthmatic? The answer is found by grasping what the immune system is programmed to do and why the hairsprings of this intricate machine respond to unwanted triggers.

The immune system is active every moment of our lives, working flat out to keep us alive when we are in serious trouble and scanning the body the rest of the time, alert to microbial intruders and all manner of irritating materials from the environment. The defenses are also alarmed by our own cells that become cancerous and purge them from the body before they do any harm. We distinguish between two arms of the immune system, although they work together.[5] The innate immune system is the first line of defense that is mobilized when immune cells recognize the chemical signatures of broad categories of invaders, namely, viruses, bacteria, amoebas, and fungi. There is very little specificity here. The body senses that it is under attack and throws the kitchen sink at these intruders. Cells that detect the unwelcome arrivals release little proteins called cytokines that summon multiple kinds of immune cells to destroy them. More bespoke defenses against specific pathogens are furnished by the adaptive immune system, which uses antibodies to neutralize infectious microorganisms.

Fungal allergies and fungal infections are very different illnesses. Fungal asthma and other allergies happen when the body is responding to mere contact with spores rather than trying to stop a fungus from growing in our tissues. The symptoms of allergy are produced by the adaptive immune system.[6] Many of the spores flowing through the nostrils get trapped on the nose hairs and in the mucus that lines the upper airways. Those that escape these obstacles float all the way down into the lungs. Proteins attached to the surface of the spores dissolve in the lung mucus. These proteins are the allergens recognized by the bodies of people sensitized to the spores, meaning that they bind immediately to the surface of cells of the immune system called mast cells and basophils. This is a chemical reaction, like a key fitting in a lock—the lock installed on the mast cells and basophils—which is perfectly shaped to receive the protein key carried by the fungus. When the key is turned, the immune cells release histamine and other molecules that cause blood vessels to dilate and the airways in the lung to constrict and fill with mucus. This is what we mean by inflammation of the lungs. Asthma is an allergy that can be caused by sensitivity to many irritants including pollen grains, dust mites, and pet dander, in addition to fungal spores. Hay fever or allergic rhinitis is another type of allergy that can be caused by spores.

FUNGI AS ONE OF THE PRINCIPAL CAUSES OF ASTHMA

Asthma was an inexplicable ailment until the twentieth century. The confusion in the Victorian era is evident in a booklet titled *Spasmodic Asthma*, published in 1879, which suggested that "atmospheric electricity" was a significant cause of asthma, along with a variety of "vegetable emanations . . . also the smell of certain animals . . . [and] dust of all sorts."[7] This was written by William Steavenson, a London physician and asthma sufferer, who discussed an assortment of treatments, including tobacco, hallucinogenic plants, and amyl nitrite (known in recreational settings today as "poppers"), and concluded, "I hardly want any other remedy so long as I have my syringe and solution of morphia." With unlimited access to his chosen narcotic, he eschewed the therapy recommended by a German professor, "who relieves his attacks by placing himself on a stool with glass legs and connecting himself with an electric machine which is worked until he is able to emit sparks from the ends of his fingers." Reading Steavenson, one gets the impression of an asthma enthusiast, someone who reveled in the study of his condition, although it is worth mentioning that lung inflammation killed him at the age of forty-one.[8]

Proof that fungi cause allergies comes from tests in which extracts from spores are pricked into the skin, and irritated mast cells release histamine causing inflammation. This immune response produces a pale bump surrounded by reddened skin and is known as the wheal-and-flare reaction. What happens on the skin is an imperfect guide to the types of allergens responsible for inflammation of the lungs, but this test is the next best thing, when inhaling different dusts to see which ones elicit an asthma attack carries the risk of death. Although self-experimentation with allergens is a dangerous venture, this was the approach taken by Charles Blackley, a Manchester physician, who provoked his own symptoms of hay fever by deliberately inhaling spores from moldy straw in the 1870s.[9] Blackley showed that spores or something else in the decomposing straw produced an allergic response. The difference between hay fever and asthma, or hay asthma, was unclear in his time, and some physicians used the terms interchangeably. Today, hay fever is the popular term for allergic rhinitis, which is the nasal allergy caused by pollen released from crop plants and by fungal spores.

There are three lines of evidence that the inhalation of fungal spores is one of the principal causes of asthma.[10] First, studies have shown high rates of fungal sensitivity in skin tests among children with severe asthma, compared, of course, with non-asthmatic controls. Next, asthma attacks and asthma deaths increase on days when airborne spore counts rise above one thousand spores per cubic meter; and third, hospital admissions for asthma increase after a thunderstorm. Thunderstorm asthma is a complicated business. It has been assumed that spore numbers increase because heavy rainfall soaks the surface of plants and stimulates the growth of microscopic yeasts and molds, and the accompanying wind gusts drive the spores of these active fungi into the air. This is the grow and blow model of dispersal.[11] But when we compare detailed meteorological records with spore counts, it appears that there is a sharp increase in the number of spores in the hours just *before* a storm.[12] This is affirmed by many asthmatics who say that they can forecast thunderstorms from a surge in their breathing difficulties, which suggests that there is more to this dispersal mechanism than high windspeeds. Adapting the famous aphorism of H. L. Mencken about answers to human problems, we can agree, "There is always a well-known solution to every [mycological] question—neat, plausible, and wrong."

Despite this evidence, fungi seem to be an afterthought for many specialists in the study and treatment of asthma. Barring the recruitment of volunteers to sit in spore-filled wind tunnels and waiting for the asthmatics to start gasping for breath, there is nothing else that can be done to convince the skeptics. Asthma is certainly caused by other irritants, but with millions of tons of fungal spores flying around the planet, the case is pretty tight. Pollen and hay fever are united in popular thinking, but mold spores and asthma remain separated. Spores are not even mentioned in *Asthma: The Biography*, authored by Mark Jackson in 2009, which is an equivalent omission to ignoring bullets in a book about gunshot wounds, or cigarettes in a study of lung cancer.[13] Jackson is not alone in the omission of the fungal connection in asthma. Even some pulmonologists (lung specialists) show little interest in the evidence that mold spores are a serious problem for many asthmatics. They have one foot in the twenty-first century and the other in the nineteenth. Clinical studies on asthma ignore the fungi, and too many physicians continue to endorse the long-standing claim that it is a psychosomatic condition, which it is not. This notion is a holdover from the era when allergies became associated with the educated classes, or "persons of cultivation," as one Harley Street physician put it in the 1880s, before he added that these afflictions were "proof of our [British] superiority to other races."[14] In the following century, a German doctor pronounced that the typical allergic patient was a delicate, "lower middle class" child, "ill-equipped for life and . . . liable to maladjustment."[15] So, asthma was, simultaneously, a badge of refinement and of ruination!

Some of the studies that have identified psychiatric contributions to asthma have ignored the challenge of disentangling cause from effect.[16] If asthmatics display anxiety-related disorders more frequently than non-asthmatics, this could be explained by the stress caused by their experiences of the illness and the resulting fearfulness of the invisible "carpet monsters," as I put it in an earlier book, that cause their lungs to shut down.[17] It is possible, too, that genes that increase susceptibility to asthma are linked to other stress responses. If, for example, asthma patients were more likely to develop depressive disorders, this would not relegate asthma to the lower status of a psychiatric rather than a physical illness, which is, of course, a false and damaging distinction in the first place. The tendency to dismiss idiopathic conditions—those without a known cause—as *merely* psychological in origin is quite widespread. Epilepsy, fibromyalgia, irritable bowel syndrome, and long COVID and other chronic illnesses following viral infections are examples of health conditions for which we have been unable to pinpoint a physical cause and have tended therefore to dismiss as psychosomatic.[18]

Marcel Proust, the most famous asthmatic, was frustrated by his father, who regarded his son's breathlessness as a big pretense that was "due to his insecure, sensitive, and dependent personality."[19] Proust captured the tragic nature of this interaction between family and invalid: "the poor suffocating patient who, through eyes filled with tears, smiles at the people who are sympathizing without being able to help him."

WHY DOES FUNGAL ASTHMA EXIST?

Allergies are caused by an oversensitivity to substances that the body treats as a threat to survival, when, in fact, the real danger lies in the symptoms of the allergy rather than the irritant itself. The troublesome proteins on the fungal spores are part of their structure and include enzymes that the fungus uses to grow in the soil and on plant surfaces.[20] They are harmless unless we react to them, which begs a question: Why is anyone allergic to fungal spores?

Compelling answers come from the perspective of evolutionary medicine, which maintains that many illnesses are rooted in the deep ancestry of *Homo sapiens* as well as the more recent history of our species. My favorite idea comes from the suggestion that the allergic response in the lungs evolved as a protective mechanism to limit exposure to noxious chemicals and to fungi that can cause lethal infections.[21] By narrowing the airways and reducing lung capacity, symptoms of asthma certainly reduce the volume of inhaled air, which would be a good thing if this was not accompanied by suffocation. But as long as the symptoms of inflammation are short-lived, the cost of decreased respiration may be worthwhile. Evolution is blind to suffering if the organism lives long enough to send its genes down the great stream of time. We have no idea how often our allergic reactions to fungal spores save us from serious infections. Many of the spores that reach the lungs belong to fungi that are very unlikely to grow in our tissues, although some show this capability in patients with impaired immune systems, as we will see later in this chapter. It is certainly possible that asthma and asymptomatic reactions to spores are lifesavers.

The risk of inhaling spores that can cause an infection has always been around because we live on a very fungal planet. The situation worsened when we abandoned our ancient hunter-gatherer and nomadic lifestyles in favor of living in agricultural settlements. Cereal agriculture demands grain storage, and stored grain is easily spoiled by molds, whose spores can become airborne in huge numbers.[22] We see the same phenomenon in cattle barns, where molds proliferate on animal feed and bedding materials and create clouds of millions of spores per cubic meter of air when they are disturbed by the livestock or farm workers. This agricultural explanation of asthma suggests that a symptomless reaction to spores and pollen that we possessed earlier in human history ramped up when we began to be exposed to masses of spores on farms. As long as the children of farmers were not debilitated by asthma and became parents themselves, the genes that controlled these relatively mild reactions to spores would have spread. The perpetuation of these forms of allergy as a protection against infection would have outweighed the costs.

The jump from relatively innocuous immune responses to spores to severe asthma may have occurred when families migrated from smaller agricultural settlements to cities, where interactions with fungi and other allergens were limited by the relative cleanliness of their homes. This seems counterintuitive, because urban life separated us from the muck of the farm, but the cleaner air meant that the immune systems of children in cities was not conditioned properly. In the first weeks of life, the infant body was not taught to ignore, or to react very gently, to the moldiness that had been unavoidable on farms. This led to asthma attacks when children with this exaggerated sensitivity were exposed to high levels of spores outdoors at certain times of the year. The problem has become heightened among children in the modern indoor environment, where severe asthma is an epidemic in some cities.[23] This explanation for asthma and other allergies is known as the hygiene hypothesis. Children who live with pet dogs and cats seem to gain some protection against allergies from the early exposure to their dander, which supports the overall virtue of training the immune system as soon as possible to deal with the rest of nature. The hygiene hypothesis remains controversial, and childhood asthma is complicated by genetic predisposition and may be worsened by bacterial and viral infections early in life.[24]

Asthmatic or not, there is no escaping the fungi. All homes are moldy. Some are very moldy. Molds grow on indoor plants and damp plant pots, and spoil fruits and vegetables in kitchens. The numbers of spores produced by fungi can become hazardous in the perpetual dampness of some older houses and in any building that is soaked by a plumbing leak, damaged roof, or flooding.[25] Encouraged by warm weather and poor airflow, fungi will feed on the natural plant-based materials in carpeting, in furniture, and on the paper that covers drywall (or plaster as it is known in the United Kingdom). The walls of some flooded homes are blackened with spores, and the chairs and couches become covered with a thick felt of spores. In the worst cases, the numbers of spores rise well above the threshold of ten thousand spores per cubic meter of air that can be very problematic for asthmatics. Even the cleanest homes blossom with mold colonies and brim with their spores. There is even some evidence that a greater variety of molds grow in homes that have been scrubbed with cleaning products that kill bacteria, which mirrors the overgrowth of yeasts on the body when we take antibiotics.[26]

There is a lot of overlap between the fungi that cause asthma in the urban environment and the most prevalent species on farms. These are the typical species that grow on all kinds of plant materials and include *Aspergillus*, *Alternaria*, *Penicillium*, and *Cladosporium*—it is very likely that you have been inhaling some of these spores since you began reading this chapter. They are, as I have said, everywhere. A few of these fungi are capable of producing harmful compounds called mycotoxins, but there is no evidence that they can reach the lungs in sufficient quantities to cause tissue damage (we examine mycotoxins in chapter 8). The problem with indoor molds is the same as outdoor molds and lies with allergy.

TREATING ASTHMA

Reducing exposure to allergens may be the best way to prevent asthma attacks, but this is difficult or impossible if we are unsure about the identity of the irritants. Vacuuming bedrooms and covering mattresses have been recommended to reduce the inhalation of proteins present in the feces of dust mites, but these methods do not turn out to be very helpful.[27] The ubiquity of face masks during the COVID pandemic provided a global test for their effectiveness in limiting asthma symptoms, but we missed the opportunity to gather data from patients. There was also some resistance among asthmatics to wearing face masks because some types produce a small dip in oxygen levels in the bloodstream, which is problematic for patients whose lung function is already impaired. Moderate improvements in asthma control have been reported in Japanese children who wear masks during sleep, whereas almost half of the American adults with asthma who responded to an online survey said that masks increased their breathing difficulties.[28] There is an opportunity here for the invention of a mask that traps fungal spores without reducing oxygen levels, but this may require a motorized pump to increase the airflow through the filter. With the discomfort and social stigma associated with wearing the simplest cloth masks, a rubbery helmet that makes a whirring sound is not going to cut it. Pending technological advances, asthmatics might consider experimenting with conventional masks during the moldiest times of the year.

Treatments for asthma symptoms are the same whether fungal spores are the trigger or pollen grains, pet dander, or dust mites. These range from drugs that counter the effects of the allergens by dilating the airways in the lungs to steroids that dampen the activity of the immune system and asthma-specific drugs that block the explosive response of mast cells when they latch on to the troublesome allergens.[29] The first of the targeted treatments was discovered by Roger Altounyan, a Syrian-born British physician and pharmacologist who suffered from severe eczema and asthma.[30] In the 1960s, Altounyan studied the effects of drugs based on a chemical isolated from a plant called bishop's weed that had been used as a folk medicine to treat asthma for thousands of years in the Mediterranean. His colleagues at a drug company had manufactured hundreds of different compounds related to the medicine from the plant, and Altounyan adopted the role of the laboratory guinea pig, inducing his own asthma attacks by inhaling dust particles and seeing which of the chemicals alleviated his breathlessness. He carried out three thousand tests over eight years. On some occasions, Roger reduced his lung capacity by 90 percent and had to inject himself with an emergency medicine to avoid asphyxiation. (It takes an asthmatic to appreciate his bravery.)

Following a eureka moment in 1963, Roger singled out a compound called cromolyn as the miracle cure. This was marketed as an inhalable medicine in 1968, in time to rescue me from a bedridden childhood. Altounyan also devised the "spinhaler" that was used to deliver the drug.

The spinhaler was fitted with a propeller that began spinning when the user drew air through the intake by inhaling. Airflow through the spinhaler distributed cromolyn powder from a disposable capsule. It is still in use today. Inspiration for the propeller came from Roger's service as a flight instructor in the Royal Air Force during the Second World War. Roger was motivated by his own asthma, balking at the prevailing medical opinion that his illness was a sign of emotional inadequacy. He died in 1987 at the age of sixty-five from an asthma attack. He is my hero.

Asthma medicines have diversified and strengthened in recent decades. This is a very good thing because more than 300 million people suffer from asthma, and the number is expected to exceed 400 million within the next few years.[31] These statistics are based on physician-diagnosed cases, and the number of asthmatics doubles to almost 700 million when we consider online responses from people who say that they have experienced wheezing. In individual countries, the asthma rates range from a low of one in fifty people in China to one in three in Australia. There is a general trend toward higher case numbers in wealthier countries, but there are plenty of exceptions. Despite their twofold difference in per capita GDP, New Zealand and Costa Rica have the highest rates of asthma in the world. Variations in the distribution of particularly irksome fungi could explain the geography of asthma, although many other factors could also influence the prevalence of the illness. By whatever mechanism, moving to a different region can prove a matchless remedy for some asthmatics. This worked for me, with relief from my breathlessness found by crossing the Atlantic, and its resumption on return visits to Oxfordshire. Mine is a compelling experiment with a sample size of me, but the geographical escape mechanism seems to be a common experience for voluntary and involuntary migrants.[32]

Allergic rhinitis affects as many people as asthma and has the same underlying immunological mechanism involving histamine release from mast cells.[33] Inflammation of the nasal passages can spread to the lower airways, and asthmatics are often plagued by both conditions. Inhalation of fungal spores can also cause a different illness called hypersensitivity pneumonitis. Symptoms of pneumonitis include breathing difficulties, coughing, and fatigue, in the chronic form of the illness, and flu-like symptoms in an acute response to inhaling huge numbers of spores or other irritants. The immune response is quite different from the inflammation of asthma and is similar to the process that underlies rheumatoid arthritis. Farm workers are frequent victims, which is not surprising, and musicians who play bagpipes and other wind instruments are also vulnerable to this condition.[34] Spores are a hazard on farms when they overwhelm the lungs of a worker moving rotting grain or animal feed and bedding, and become a problem in bagpipes, trombones, saxophones, and tenor horns when moisture and phlegm from the players combine with the interior coatings of the instruments to create a matchless breeding ground for fungi. Mushroom workers are also prone to hypersensitivity pneumonitis for the more obvious reason that their livelihood depends on cultivating millions of natural spore fountains in enclosed spaces.

The need for research on fungal allergies and the development of effective medicines for alleviating the worst symptoms is growing because climate change is likely to increase the number of spores in the air. Regions that experience warmer and wetter weather will become especially moldy as fungi flourishing on plant debris generate more spores. This asthmatic future is developing already, with a major study from the San Francisco Bay Area showing that the mold and pollen seasons have been extended every year since 2002, increasing the number of days when asthmatics are inhaling lots of spores and pollen.[35] Long-term trends in spore counts are very difficult to predict and will respond to regional differences in weather patterns and changes in land use, including the clearing of grasslands and forests for cereal cultivation. Fellow asthmatics: keep your inhalers ready.

FROM ALLERGY TO INFECTION

Fungi that cause asthma and other allergic illnesses are visitors to the human body, fly-by-nights rather than long-term residents of the mycobiome. Some of the spores that carry allergens can also germinate in the nasal passages and the lungs and develop into mycoses, but the healthy body sheds these along with every other spore and irritating dust particle using mucus as a trap. With each inhalation, air is vacuumed through the 12-millimeter-wide windpipe, which divides into the pair of bronchi, and onward into the narrower airways that branch more than twenty times until they feed into the alveoli. The alveoli are attached to the tiniest branches in the lungs and look like bunches of grapes. This glistening labyrinth has a surface area of 100 square meters, which is three or four times larger than the skin and its hair follicles. Spores and bacteria and other particles stick to the mucus that lines the breathing apparatus, and this viscous fluid is swept upward by trillions of cilia attached to the cells lining the tubing that lash around ten times per second.[36] This motion clears the mucus from the microscopic alveoli, all the way up to the trachea, or windpipe, between the vocal cords, and into the back of the throat. Mucus reaching the throat is swallowed and dissolves in the acid bath of the stomach. This mucus conveyor belt runs continuously, transporting fluid all the way from the alveoli to the throat in about six hours. Coughing accelerates the waste-removal system by forcing gobbets of mucus from the lungs into the throat and spraying thousands of droplets into the air at a speed of 60 miles per hour (100 kilometers per hour). Sneezes are even more violent, discharging one hundred thousand drops of mucus at twice the speed of a cough.

We are unconscious of this brilliant machine until it gets clogged, and fungal infections tend to develop when this happens. Cystic fibrosis is an illness that messes with the conveyer belt by thickening the mucus. This means that fungal spores and bacteria are not cleared as swiftly as normal, and the condition leaves patients open to infection than those with sufficient fortune to be born with runnier mucus. Spores of *Aspergillus* species are the most problematic because they are everywhere and can germinate if they are not cleared from the lungs. This can lead to allergic bronchopulmonary aspergillosis (ABPA), which is an illness that worsens mucus buildup and obstruction of the airways in cystic fibrosis patients.[37] ABPA also occurs in asthmatics whose illness does not respond to the usual medicines. Fungi can also colonize the lungs and develop into serious infections. Antifungal drugs can be effective at eliminating these fungi in the short term, but this does not reduce the risk of a later infection. Any of the health conditions that diminish lung function increase the likelihood that fungi will settle in the recesses of our airways and implant themselves in our tissues.

Aspergillus species and other fungi get stuck in the nasal sinuses too and can expand into a clump of filaments called a fungus ball that presses on the surrounding tissues. Symptoms range from headaches to pain and tenderness in the area of the sinuses above and below the eyes and on to changes in vision and proptosis, when the eyeballs protrude from their sockets. Long before most patients get to this stage, surgery is performed to remove these hideous fungal excrescences. Besides aspergillosis, the most prevalent fungal lung infections are histoplasmosis, blastomycosis, and coccidioidomycosis, whose names refer to the fungi that cause them.[38] Fungal growth in the lungs is countered by the cells of the innate immune system that hang around in the lungs, primed for action. These are macrophages and neutrophils that gobble up the germinating spores and release inflammatory chemicals to recruit more cells to the battle. Neutrophils are the most abundant type of white blood cell, with twenty billion or so circulating in the bloodstream, which equates to fifty thousand in a pinpricked drop. Predatory neutrophils are larger than the spores that they consume, but the feat of engulfing one is comparable to a human swallowing a Halloween pumpkin. Neutrophils also deploy chemical weapons that work as disinfectants and bleaching agents to destroy fungi without having to eat them.

When the number of neutrophils drops below a critical threshold, the fungi overcome these defenses and penetrate the walls of the lungs and move into surrounding tissues. Fungi can also make their way into the bloodstream and become distributed all over the body in disseminated infections. This explains why illnesses that reduce neutrophil counts, including leukemia, anemia, and HIV/AIDS, are associated with aspergillosis and other lung infections.[39] The risk of serious mycoses also increases when the immune system is upset by steroid medicines used to support patients after organ transplantation and to treat allergies and autoimmune diseases.[40] Cases of aspergillosis increased during the COVID pandemic when patients who developed pneumonia were treated with these drugs to control lung inflammation.[41]

Aspergillosis is a global infection, whereas the fungi responsible for histoplasmosis, blastomycosis, and coccidioidomycosis are confined to particular regions.[42] Histoplasmosis, caused by *Histoplasma*, lives in the central United States. This mycosis is also known as Ohio Valley disease. The fungus thrives in bird and bat feces, and so its spores are concentrated in chicken coops, caves, and abandoned buildings. Everyone in this region is exposed to the fungus, but its growth in the body is stopped in its tracks by a functioning immune system. The distribution of *Blastomyces*, which causes blastomycosis, overlaps with *Histoplasma*, and extends northward around the Great Lakes and southward to the Gulf Coast. It is a soil fungus and is destroyed by macrophages and neutrophils in the lungs. Coccidioidomycosis, or San Joaquin Valley fever, is caused by *Coccidioides*, which grows in the southwestern United States and northern Mexico and is also established in parts of South America. It is a soil fungus whose spores are lifted into the air in dust storms. Like the other mycoses, few people show any symptoms of infection by this unpronounceable fungus before it is squelched by the immune system. Dogs are also susceptible to this fungus and suffer from higher rates of infection and more frequent complications.[43] Related species of fungi are responsible for an African version of histoplasmosis and South American variety of coccidioidomycosis in humans.

None of these mycoses that begin as lung illnesses are common compared with the burden of viral infections. Aspergillosis affects 300,000 people globally every year; the American and African species of *Histoplasma* cause 100,000 annual infections; 25,000 patients develop coccidioidomycosis; and blastomycosis is the rarest of the quartet, limited to 3,000 infections in the eastern United States.[44] Regional outbreaks of these diseases occur every year. Wisconsin is a hotspot for blastomycosis, where humans and dogs are attacked by the fungus that grows in soils along riverbanks and rotting vegetation. In the adjoining state of Michigan, more than ninety workers developed the illness in a paper mill in 2023.[45] These disease clusters attract media attention and encourage the impression that fungal lung infections are becoming more common, but the evidence is equivocal. Either way, the most serious symptoms of fungal infection develop when the immune system is in very bad shape. Antifungal drugs can be effective at halting the growth of the fungi or eliminating them completely in some patients, but the loss of those neutrophils means that the infections are liable to reboot from residues of the fungi in the body or from fresh spores arriving in the lungs. The mortality rate for these pulmonary mycoses soars in older patients whose immune systems have collapsed.

All of these lung infections begin with the inhalation of spores that are floating in the air. One infectious fungus behaves very differently. This is *Pneumocystis*, a yeast that does not produce spores at all. *Pneumocystis* is found in the lungs without causing any illness, but develops into a full-blown, life-threatening form of pneumonia in HIV-positive patients when they develop AIDS.[46] The incidence of *Pneumocystis* pneumonia, or PCP, follows the number of AIDS cases, so it is not surprising that this mycosis is common in Nigeria and other African countries which have the highest rates of HIV infection. PCP also develops in transplant patients. There are many unanswered questions about *Pneumocystis* because it cannot be grown in the laboratory in culture dishes. We are not even sure how it gets into the lungs, although outbreaks of the infection in hospitals suggest that the yeast cells of the fungus are transmitted from person to person in droplets of lung mucus expelled by coughing or normal breathing. No other fungus seems to be dispersed in this viral fashion.

Our breathing apparatus is a glorious contraption when it works perfectly. Hindus believe that we are allotted a predetermined number of breaths for our lifetime, which approaches five hundred million ins and outs for the global average life expectancy of seventy-three years. With half a billion flushes of gas over the 100 square meters of tubing and air sacs in our bodies, this is the region of greatest contact with the fungi and affords the greatest prospects for infection. Allergies to spores and the opportunities for fungi to take root in the lungs fall into the category of unpleasant interactions with the mycological world. There is a lot of uplifting news to come in this book, but next we will look at how bad things can get when fungi spread around the body. Even Proust would agree that these deep-seated mycoses make an asthma attack seem like *une promenade de santé.*

4

Spreading

OPPORTUNISTS IN THE BRAIN

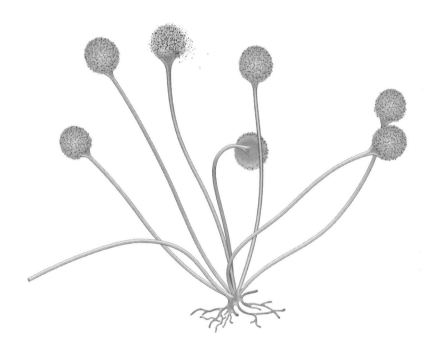

IN 1997, fifteen-year-old tennis player Sasha Elterman was thrown into the polluted water of the Yarkon River in Israel when a footbridge constructed for the Maccabiah Games collapsed. The Maccabiah Games are held every four years and are often referred to as the Jewish Olympics. Sasha was a member of the Australian athletics team, which had been crossing the bridge when the support beams broke. Struggling to keep her head above the surface, she swallowed and snorted the filthy water before she was dragged to the muddy bank. One athlete died at the scene, and sixty-seven team members were taken to local hospitals. None of the injuries seemed to be life-threatening at first, but within hours many of the victims, including Sasha, developed breathing difficulties. Three more athletes died over the next few weeks. Early on, it was thought that they were poisoned by chemical pollutants in the river, which included oil, solvents, and toxic heavy metals. These ideas were scrapped a few days later when analysis of lung tissue from one of the autopsies revealed something else: the threads of a fungus called *Scedosporium*.

Sasha was flown to a hospital in Sydney after treatment in Israel and remained in critical condition for months. She continued to be upbeat during this ordeal, but the development of a brain infection by the fungus worsened her prognosis. As the illness progressed, she endured multiple surgeries to remove infected tissue from her damaged lungs and brain.[1] All the while, the fungus resisted a battery of drug treatments. It kept coming back. Running out of options, her doctors decided to try a new medicine called voriconazole, developed by Pfizer in the United Kingdom. As this potent compound suffused the infected tissues, the microbe began to lose the fight, disappearing in one location, then another, until, at last, the ghastly fungus was gone. Three years later, after intensive rehabilitation, Sasha was well enough to carry the Olympic torch in the relay at the Opening Ceremony of the 2000 Summer Olympic Games in Sydney.

Sasha's story tells us something about a very different kind of relationship between humans and fungi from the lifelong interactions between the body and the resident mycobiome. Most of the serious mycoses are caused by fungi that we encounter in the environment every day but that cause infections only when the immune system is debilitated. Sasha's fungus was an oddity that was forced into her body by her accident and battered its way through her intact immune defenses. To understand this case history, we need to look at the fungus from the Yarkon River more carefully. Like most of the fungi in this book, *Scedosporium* has never been given a common name. We are stuck with the Latin, which refers to seeds or spores. This fails to conjure up a distinctive image of the microbe, so imagine looking through a microscope: within the bright circle, the fungus comes into sharp focus as a gossamer of cylindrical filaments, plumped with internal droplets of oil and bristling with stubby side branches bearing oval spores. The fungus is beautiful in its simplicity when it is grown in a culture dish in an incubator; sinister when highlighted with purple stain in a thin slice of brain tissue taken from a refrigerated cadaver in the morgue.

Fungal brain infections in people who come close to drowning are rare and very difficult to treat when they are diagnosed. A case of *Scedosporium* infection in Germany involved a forty-one-year-old woman in an automobile accident who was thrown from her car into muddy water. She was resuscitated at the scene but developed several brain abscesses over the next five days. These appeared as white spots on her MRI scans, some as big as grapes, and the fungus was identified as the culprit from DNA extracted from samples of brain tissue. The woman experienced a range of distressing neurological problems including epileptic convulsions as her doctors exhausted the catalog of antifungal drugs. Her condition improved, eventually, and her case history was published after ten years of treatment.[2] At that time, she had been in a stable condition for two years, which speaks to remarkable resilience. A study of more than a hundred cases of infection by this fungus revealed a median survival time of just four months.

The mechanism that *Scedosporium* uses to invade the brain has not been solved, but the central nervous system is an attractive sanctuary from the point of view of the fungus. Tipping the scales at an average of 1.3 kilograms, the brain contains more calories than a large roast chicken.[3] Getting in there is difficult. The hard skull is an obstacle for animal predators, which explains why lions and other carnivores go for the more accessible organs slithering from the abdomen when they rip into their prey. Microbes make their way into the brain via blood vessels that snake through perforations in the skull called foramina and fissures. This gets them into the head, but before they can reach the squishy nerve tissue, they have to overcome the blood-brain barrier, which is a formidable challenge. This is formed by a layer of cells that lines the interior of the blood vessels feeding the brain. These cells are wedged together, side to side, preventing large molecules and any microorganisms circulating in the bloodstream from passing into the brain.

Fungi use a number of different mechanisms for crossing this hurdle. Some mount a chemical attack, releasing enzymes that weaken the protective cell layer. Others use a Trojan horse strategy, hiding inside white blood cells that move across the barrier as part of the natural operation of the immune system and spilling out on the other side. Once they find themselves brain-side, the fungal cells multiply, forming the abscesses that show up as islands of damaged tissue on CT and MRI scans. The immune defenses are concentrated in the outermost tissues of the brain, where they resist invaders. Once inside the brain, the fungus is free to do its worst.

The association between brain infection and traumatic incidents of near drowning suggest that the blood-brain barrier is weakened by the pressure of the water forced into the nasal passages. Carried with the water, the fungus responds by trying to hang on, doing everything it can to survive. In Sasha's case, and the German patient, the fungus found food, grew itself in knots, and spotted the brain with abscesses. This species does not seem to be especially adapted to growing in brain tissue, but it does a lot of damage when it finds itself there.

Scedosporium does not confine itself to victims of near drowning. It also multiplies in patients whose immune systems have been weakened by HIV infection or cancer, and in those whose defenses are lowered by drug treatments following organ transplantation.[4] The loss of immune function makes it more likely for a fungus to spread into the bloodstream. *Scedosporium* is widely distributed in nature and seems to be particularly prevalent in water contaminated with sewage from humans or farm animals. Industrial farms create a perfect habitat for this fungus in the form of the abundance of animal waste dammed in ponds lined with black plastic and weighted down with tractor tires around the rim. One spectacularly unpleasant case history involved a young man in Brazil who died from a brain infection caused by the fungus three months after he fell into one of these reservoirs of swine sewage.[5] *Scedosporium* also lives in lake water and has been found in the soil of potted plants in hospitals. The fact that so few people become infected with this fungus says everything about the power of the immune system to protect us, as long as we avoid a rare accident and nearly drown in water that it calls home.

OPPORTUNISTS

The fungus that sickened Sasha Elterman is one of thirty or more species that have been identified in brain infections, and these are a subset of the three hundred kinds of pathogenic fungi that cause disease all over the body. Given that there are more than seventy thousand species of fungi, and some experts think there may be more than one million, the nasty ones belong to a tiny minority—less than 1 percent of the total number of species that have been described by scientists and given a Latin name.[6] The pathogens represent a splinter group from the great fungal kingdom, whose principal concern over hundreds of millions of years has been with decomposing dead plants and partnering with live ones or attacking them. Doing the same things with animals—rotting, cohabiting, and infecting—is a secondary profession for the mycological world. Next to these long-standing activities, making our lives a misery is a very recent specialty. Because humans have such a short evolutionary history, the fungi that invade human tissues were occupied with other tasks long before they found themselves inside our bodies. This explains why, by and large, they are not very good at making their own way from the outside environment into our tissues. Even though they are total losers as pathogens compared with viruses, they still cause a lot of trouble, killing more than 1.5 million people every year. This is an astonishing toll when we consider that *only* four hundred thousand people die from malaria.[7]

Mortality figures for fungal illnesses, meaning how many infected people die, match those for tuberculosis, which is caused by a bacterium. Many of the deaths due to tuberculosis and to fungal disease occur in AIDS patients whose immune defenses have been overwhelmed by HIV infection. Physicians who treated AIDS cases in the early 1980s, before the virus was identified as the cause of the illness, were alarmed by a surge of fungal infections seen in young men. Patients displayed a form of fungal pneumonia as their immune systems failed and fungal brain infections became another sign that someone had developed full-blown AIDS.[8] Serious fungal infections are much less frequent in HIV-positive patients today if they are receiving the excellent drug therapies that control the virus, but proper treatments are scarce in parts of sub-Saharan Africa and Southeast Asia.[9]

Research on the link between AIDS and fungal infections has helped to explain how the functioning immune system keeps the body free from these diseases. The greatest damage from the virus comes from its destruction of a specific type of white blood cell that is a key player in the seek-and-destroy mission of the immune system. These are the helper T cells. This also explains why certain forms of leukemia that deplete these cells are associated with the same mycoses. A similar reduction in white blood cell count is seen in patients treated for cancer by chemotherapy or radiation therapy, as well as in transplant recipients who take drugs to prevent organ rejection. When the shield of T cells fails, the onboard mycobiome becomes restless, mottling the skin, plugging the nasal sinuses, whitening the tongue, and fouling the throat before spreading from the lungs and the gut to the liver, kidneys, and brain. These harmless symbionts that turn bad are joined by airborne spores that land on the defenseless body, and we are taken apart piece by piece. The fungi that drop anchor after immunological damage or injury are called opportunists or opportunistic pathogens. All of the fungi that cause serious infections in humans are opportunists. Although only a few hundred species of fungi have been associated with tissue damage, it is possible that thousands of fungal species can harm us if they find themselves in a defenseless body. It has even been suggested that the ability to cause disease in humans is a defining characteristic of the kingdom.[10]

This concept of universal pathogenicity seems ridiculous when we think about mushrooms that grow in the woods, but colonies of these fungi that form fruit bodies do cause lethal infections.[11] Human tissues are not the preferred food for mushroom mycelia, but these fungi make do when they find themselves in an unprotected body. Consider the case of a six-year-old girl with kidney cancer who developed a swelling on her head that split open and discharged pus. When samples from the wound were transferred to a culture dish, the pathologists were shocked by the growth of a mycelium of an ink cap mushroom that lives on animal dung in the wild.[12] The girl was treated successfully by surgery to remove the infected tissue and a course of antifungal medicines. This was a bizarre infection, although the same mushroom has been found in lung tissue and can damage heart valves after cardiac surgery. Authors of a case history from the Mayo clinic involving a seventy-seven-year-old woman with clots on her replacement mitral valve titled their report, "Truffle's Revenge: A Pig-Eating Fungus."[13] The ink cap mushroom had grown over the "bioprosthetic" valves, which had come from a pig.

Appearing less menacing than any mushroom—indeed, as harmless as a loaf of bread—*Saccharomyces cerevisiae*, the yeast used for raising dough and brewing beer, causes lethal infections in exceptional instances when it passes into blood vessels through a catheter. The idea that the fungus purchased as a freeze-dried powder in the grocery store can kill seems absurd. But it can, and, like the ink cap, baker's yeast is a perfect example of an opportunist.[14] It is important to recognize that these are extreme curiosities in the literature of infectious disease that should not discourage mushroom hunting or alarm any bakers or brewers. These freakish infections are astonishingly rare.

The best way to think about opportunistic pathogens is that they sit along a whole spectrum of behavior, with varying degrees of preparedness and capabilities for messing up our lives.[15] Fungi that cause athlete's foot and toenail infections are examples of more purposed pathogens than the ink cap mushroom because they are so well adapted to growing on skin and nails. These fungi live parallel lives in soil, where they consume scraps of animal protein and other organic materials, but they are accomplished at making themselves comfortable when we pick them up by walking barefoot over their territory. Contact with these fungi does not necessarily lead to an infection, however, because some people are affected by athlete's foot throughout their lives and others are not.

Returning to the fungi that grow in the brain, they share some characteristics that are fitted to this loathsome business.[16] The ability to grow at the elevated temperature inside the body is an obvious prerequisite for a fungus that causes brain infections. This is not asking much of microorganisms that thrive in the summer temperatures experienced over much of the planet, but it does discount species adapted to cooler climates. Fungal pathogens must also be equipped to outwit the remaining strength of the immune system in weakened hosts. A lot of the brain pathogens appear to benefit from the presence of melanin within the walls of their cells that gives them a black or brown color.[17] This fungal version of melanin is a different pigment from the chemical that colors human skin, and it acts as a chemical mop that neutralizes some of the natural disinfectants produced by the immune system. Pigmentation may also help the fungus in other ways, by stabilizing its cells at higher temperatures and furnishing protection against ultraviolet light. Despite these features that help some fungi grow inside the body, the prevailing view of experts in medical mycology is that these opportunists do not want to be there in the first place.

To understand this reasoning, we need to think about evolution. Viruses and bacteria that cause infectious diseases multiply in our tissues and move from person to person in droplets released by breathing, sneezing, or coughing, via skin contact and through sexual behavior. Insects and other animals act as vectors that transmit viruses and bacteria, and mothers can pass infections to their developing babies through the placenta and in breast milk after birth. This list of infection pathways covers most of the ways that microbes spread between humans. Fungi that grow deep inside the body have no mechanism for escaping.[18] This means that a fungus that forms colonies in the brain is doomed. It will die with its host. If the corpse decomposes in soil, the fungus may seep into the dirt as the tissues dissolve and go on to reproduce in the environment, but there is nothing about lingering in the host that made the passage worthwhile. Infecting humans is a dead end for fungi, which explains why they are no good at causing pandemics like viruses. Molds that cause athlete's foot are an exception to this rule, and the yeast *Candida auris*, which is causing serious infections in hospital patients, does not threaten the general population (see chapter 2).[19] Most fungi are happy in the soil, and we would be happier if they stayed there. Fungal infections of humans, or mycoses, are part of the noise of biology that present no advantages to the pathogen or the host. These mutually harmful relationships have been termed synnecroses.[20]

Even though it is content growing outside the human body, another fungus called *Cryptococcus neoformans* is remarkably good at causing brain damage. It has attracted the attention of medical mycologists since it was identified as the agent of brain infections in AIDS patients. Since then, cryptococcosis has become a disease of global proportions that is responsible for more than half a million deaths per year in the developing world. The fungus spreads through the brain, damaging nerve cells and forming cysts in different regions. The membranes, or meninges, that surround the brain become inflamed, and this results in brain swelling. As the infection develops, symptoms include persistent headaches, neck pain, and drowsiness, and these can progress to disorientation, difficulty finding words, nausea and vomiting, leg paralysis, convulsions, strokes, and death. Although rare infections by this fungus occur in otherwise healthy patients, most cases of cryptococcosis are associated with weakened immune defenses, which explains why the disease is more common in countries with high rates of HIV infection.[21]

Cryptococcus is a soil fungus whose growth is energized by bird droppings. Utopia for this fungus is a chicken coop or pigeon roost, and it does not need to waste any time inside human beings. Getting into us as an airborne spore is a misstep. Most fungal spores that we inhale are swept from the narrowest airways to join the conveyer belt of mucus that moves upward to the throat and drops down into the stomach where the daily dose of microbes goes to die. *Cryptococcus* is one of the few organisms that can dodge this fate when conditions are ripe, cross into the bloodstream from the lungs, and move through the barrier into the brain.

The ability to outwit the immune system, especially if it is weakened, is probably a consequence of the natural behavior of the fungus in the soil where it grows as a form of budding yeast. These cells are preyed on by amoebas, which consume all kinds of microbes in the soil and digest them in food vacuoles within their cells. Certain strains of *Cryptococcus* avoid this fate and manage to stay alive inside the vacuoles, and the same trick allows the fungus to survive when it is engulfed by the macrophages of the immune system that feed like amoebas. (Strains are like breeds rather than separate species.) They stay inside the food vacuoles of the macrophages, hitchhiking until they are vomited, unharmed.[22] Cell biologists call this mechanism vomocytosis, so I am not being overly poetic here. The bad stuff unfolds when a macrophage with stow-away *Cryptococcus* crosses the blood-brain barrier and releases its cargo. The life of this fungus will end when the patient dies, but, in the mean-time, it feeds and reproduces by forming buds, and the brain abscesses multiply with each CT scan.

Treatment options for cryptococcosis are very limited. The handful of drugs used to combat this infection have serious side effects and have not been updated since the 1990s.[23] Amphotericin B is a natural product iso-lated from a soil bacterium. It disrupts the cell membrane of the fungus but also damages the kidneys. A second medicine, flucytosine, interferes with the formation of DNA and proteins in the fungus. The problem with this one is that it causes liver damage. Fluconazole is the third antifungal drug used to treat cryptococcosis. This belongs to the azole family of an-tifungal agents that also target the cell membranes of fungi. It has fewer side effects than the other medicines, but its drawback is that it limits the growth of the fungus without killing it. For this reason, it is used for "maintenance therapy," to keep patients in a stable condition. It cannot rid them of the infection. Someone with a strong immune system who contracts the disease can be cured with a combination of these drugs, whereas the long-term outlook for a patient with weakened defenses is not as reassur-ing. The mortality rate for cryptococcosis for HIV-positive patients ap-proaches 80 percent within one year of diagnosis in some developing countries. These disheartening statistics and the inadequate treatment options led the World Health Organization to rank *Cryptococcus* in the Critical Priority Group of pathogens that require urgent interventions, including the development of new drug therapies.[24]

THE HORROR OF MUCORMYCOSIS

Few people have ever seen *Cryptococcus* for themselves, or any of the other pathogenic fungi that I have described, for that matter. This would take a microscope and access to samples of infected tissue. You would need to have taken a microbiology course in medical mycology to have enjoyed this honor, and these are very few and far between. Surprisingly, however, there is one type of fungus that can disfigure faces and destroy brain tissue that almost everyone has seen *without* a microscope. These are the black bread molds, which are so common on spoiled tomatoes and other fruit that we do not give them a second look before relegating the rotting food to the compost heap. The mycelia of these species of *Rhizopus* and *Mucor* feed on our groceries before forming millimeter-tall translucent stalks tipped with blackened bulbs. Each of these bulbs contains hundreds or thousands of microscopic spores that are dispersed by air currents and start new colonies when they land in our fruit bowls. The stalks are quite beautiful when they are magnified with a hand lens, appearing as a miniature forest of crystalline stems bearing their tiny, blackened globes. It is difficult to reconcile this prettiness with the photographs of swollen, reddened, and sometimes noseless and eyeless faces of the victims of "mucormycosis" on the internet. But this is what these fungi can do.

The first thing to make clear is that there is nothing that we can do to avoid this infection, and that with an estimated two cases of mucormycosis per million people per year, you are more likely to be attacked by a moose than a bread mold.[25] The reason that the infection is unavoidable is that the spores of these fungi are floating around indoors and outdoors in huge numbers and they get trapped in our nostrils every day. The rarity of the infection is more difficult to explain, but there are some pointers. Many patients who develop mucormycosis have preexisting illnesses including cancer and uncontrolled diabetes or have been severely burned. Others are taking medicines that interfere with their immune systems. Cases also occur in patients after surgery, and in premature babies.

The bread molds show no subtlety in the biological mechanism of their attacks. There is none of the Trojan horse behavior seen with *Cryptococcus*, which allows the fungus to slip unnoticed into the brain hiding in a macrophage. In mucormycosis, the spores germinate in the mucus in the nasal passages, send their filaments into the soft tissues, and wangle their way into the brain by growing along the walls of the blood vessels. The disease is a classic example of an opportunistic infection. It has maintained a stubborn mortality rate above 50 percent for decades because surgery is the only treatment and involves carving away the infected tissues. This is the reason that some patients lose an eye to the disease or, worse, are left with a large opening in the middle of their face. This is as horrible an infection as one can imagine. It is the stuff of nightmares, and none of the antifungal drugs seem at all effective in controlling this beast once it takes hold.

Mark Tatum, a forty-four-year-old from Kentucky, suffered a horrific encounter with one of these fungi in 2000 that destroyed much of his face. To save his life, surgeons removed his eyes, nose, upper jaw, and masses of surrounding soft tissue and cheekbone. To counter the pain following this butchery, he was placed in a drug-induced coma for two months. His wife, Nancy, said, "His doctors told me it was one of the most extensive surgeries they'd ever performed on a person's face."[26] She continued, "When I went into the critical-care unit . . . they expected me to faint, but when I looked at my husband, I just saw Mark. . . . I looked into the cavity on his face. . . . I saw the lining of his brain and the top of his tongue, but I know Mark is more than his face. He is my husband, the man who gave me everything I had ever wanted." Something of his original appearance was restored with a removable prosthetic mask that was attached to a frame with magnets. His story was broadcast on television, and his bravery and good humor inspired people all over the world: "I didn't do nothing noble," Mark said, "I just did what was necessary." He and Nancy pursued their lives with astonishing grace until his death in 2005.

Mark Tatum's infection may have been triggered by his use of steroids to treat back pain. Corticosteroid drugs work by dampening the inflammatory response that is one of the foundations of our immune defenses, meaning that they can make us more likely to become infected by fungi. This is the reason for the epidemic of mucormycosis in India during the COVID-19 pandemic. More than four thousand "black fungus" deaths were reported by the summer of 2021 in patients treated with corticosteroids to mitigate the hyperimmune reaction to the virus known as the cytokine storm.[27] (This storm of inflammatory compounds damages the lungs and other organs and was a leading cause of death from

COVID-19.) On the plus side, steroid treatment was a lifesaver for many patients who were critically ill with COVID-19, and the number of fungal infections remains vanishingly small among the hundreds of millions of people treated with steroids for other illnesses.

Another forceful illustration of the link between steroid use and fungal infections comes from an outbreak of a different mycosis in patients given spinal injections with a corticosteroid for pain control in 2012. Across the United States, there were more than seven hundred cases of meningitis and spinal infections, and sixty-three patients died. An investigation by the Centers for Disease Control determined that the infections were caused by a fungus called *Exserohilum*, which normally grows on grasses.[28] Batches of the drug preparation were contaminated with the fungus. It would be difficult to think of a more powerful demonstration of the connection between the immune system and development of fungal disease. The injection of the fungus into the spine along with a drug designed to muzzle the immune system guaranteed disaster. This tragic accident illustrates how modern medical practices can make us vulnerable to the oddest kinds of fungal disease.

THE BRAIN MYCOBIOME

Even though the chain of events leading to most serious fungal infections is rarely as clear as the meningitis cases caused by the contaminated spinal injections, we can find answers by examining the medical records of patients before their infections, identifying the fungi growing in their tissues, and tracking the progression of their illnesses. Sometimes we can develop a pretty good picture of what happened, but other case histories remain baffling. Why one person in a million with an apparently healthy immune system develops an incurable infection by a fungus that seems to be present in just about every soil or water sample on the planet is beyond our comprehension. Whether a physician knows how their patient became infected, or has no idea, the options for treating the illness are the same and remain too limited for comfort.

This sense of uncertainty in the field of medical mycology has grown with the highly controversial claim that fungi may be involved in

Alzheimer's disease and other neurological conditions. This idea has been spurred by the identification of fungal DNA in brain tissue from Alzheimer's patients sampled at autopsy.[29] The DNA comes from a range of species, and the work is backed up with microscopic images of yeast cells and filaments in samples from different regions of the brain. Alzheimer's disease is associated with the presence of specks of misfolded proteins in the brain known as amyloid plaques. Plaques are important because they are part of the inflammatory response that is characteristic of Alzheimer's disease and are linked to the death of nerve cells. They may develop when the body's immune system begins to attack the tissues that it is supposed to protect. This is the autoimmune model for the disease. A second idea is that the inflammation and formation of plaques is a response to an infection.[30] This is supported by experiments on mice infected with *Candida* yeast, whose brains become damaged with plaque proteins.

The work on fungi in the brain is so new that no definitive conclusions can be drawn yet, and the problem for scientists is the stubborn challenge of separating cause from effect. It is possible that Alzheimer's has an underlying cause that has nothing to do with infection and that the fungi and other microbes arrive once the brain is already damaged. If the fungal connection is supported by other experiments, the next thing we need to figure out is whether the fungi migrate to the brain from elsewhere in the body or whether they come from the environment. Are brain fungi long-term residents or recent immigrants?

Some investigators have reached beyond the available data to suggest that fungi are part of a cryptic microbiome that lives inside the brain before the development of any neurological disease. In a very limited study, they have found bacteria clustered around star-shaped cells called astrocytes in samples of healthy brain tissue taken from fresh cadavers.[31] In addition to bacteria, traces of fungal DNA have been detected in these brains. *Fusarium* is the most frequent type of fungus that has been identified. Many investigators remain skeptical about these findings, and it is possible that the brain samples in these studies were contaminated after the death of their owners.

Inflammation of brain tissue features in other neurological diseases, raising interest in the possibility of a widespread fungal connection in many neurological conditions whose mechanisms have always seemed puzzling. Amyotrophic lateral sclerosis (ALS) is the dreadful disease that afflicted the physicist Steven Hawking. It is also known as Lou Gehrig's disease, after the famous baseball player who died in 1941. In a speech at Yankee Stadium on July 4, 1939, Gehrig described his illness as a "bad break" and went on to describe himself as "the luckiest man on the face of the earth," which makes me think of the bravery of Mark Tatum. Nerve cells that control voluntary muscles are destroyed in ALS, meaning that patients lose control of their conscious movements. *Candida* and other fungi have been found in the brains of ALS patients, but the question of cause or effect remains.[32] Genetics seem to be a factor in 5–10 percent of ALS cases, but the majority are described as sporadic, in the sense that they develop without any clear predisposing factor (spontaneous or idiopathic are better terms). This pattern of illness is consistent with the influence of some unidentified environmental factor, such as an infectious agent.

The brains of people who die from Parkinson's disease are also colonized by fungi.[33] DNA and cells of *Candida* and *Fusarium* show up again, along with the scalp yeast *Malassezia*, and *Botrytis*—a fungus that is a common pest of fruits and flowers. After the Canadian American actor Michael J. Fox was diagnosed with Parkinson's disease in the 1990s, there was a lot of interest in the phenomenon of disease clustering in ostensibly noninfectious illnesses. Fox's case had the unusual attribute of being one of four diagnoses of Parkinson's disease among the cast and crew of a television series that had been filmed in British Columbia in the late 1970s.[34] There are many possible explanations for a disease pattern of this kind, but exposure to a particular microorganism could play some undiscovered role in these life-changing and life-ending illnesses. We seem to be a long way from a definitive answer. The reports of fungi in brain tissue have survived the rigors of peer review to appear in excellent journals, but most of the work has come from a single group of scientists and deserves a lot more attention.

The body is such an intricate machine that its endurance, day in and day out, can seem miraculous. After all, there are so many things that can go wrong with a creature that depends on an infinitude of fine-spun biological mechanisms. On the other hand, we are the recent products of a successful evolutionary history that has compelled our twenty thousand genes and trillions of cells to work together for at least as long as it takes for us to reproduce. This much is self-evident. We would not be here at all if we were really as fragile as many of us fear. The body develops as a comfortable home for herds of microbes, and rather than surviving as individuals, we have already seen in this book how we teem with invisible life—fungal and otherwise. Far from immaculate, the body prevails as a mobile ecosystem harmonized by the brilliance of the immune system. While any fungi that live in a healthy brain seem to be rare, we find a very different situation in the digestive system, to whose more prosperous mycobiome we turn now.

5

Digestion

YEASTS IN THE GUT

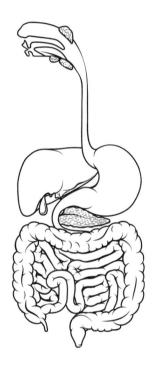

HOW WOULD YOU rate your digestive system? Does it operate like a well-oiled machine or a malodorous trash compactor? Most of us would probably say, "Somewhere in between," and add that its performance varies from day to day. An uneventful and ignorable intestine is the gold standard gut, but even the best of bowels are rattled by an ill-chosen meal. The trillions of bacteria in the microbiome of the digestive system have received a lot of attention, whereas the fungi that wax and wane in their midst have played second fiddle or been ignored—until now. New species of fungi are introduced to the body on fresh fruits and vegetables, and others are long-term residents in the gut. Some of the newcomers die in the stomach acid, and others survive downstream to make war and peace with the existing microbes in the intestine or ride within the waste until they escape from the body. The fungi are there for the whole journey from mouth to esophagus to stomach and onward to the small intestine, large intestine, rectum, and beyond. This is the richest and most mysterious part of the human-fungus symbiosis.

Until recently, the study of the fungi that affected human health was limited to the fungi that cause ringworm on the skin and life-threatening infections of our internal organs. This constituted the study of medical mycology in the twentieth century. In hospitals, mycologists who were brought in to look at cases of serious disease examined the fungi seen in microscope preparations of tissue samples and grew the fungi isolated from patients in culture dishes. These techniques enabled them to identify the fungi and advise physicians on treatment methods. Although mycologists were aware that some fungi grew in the gut, these yeasts were barely mentioned. They did not seem to be doing anything significant. The application of methods to amplify the DNA of microorganisms from samples of feces did not make much difference, at least initially, because the techniques were perfected for identifying bacteria (mentioned in chapter 1). This led to the treatment of the gut microbiome as a giant onboard bacteriome. Untangling the fungi from this assortment remains difficult.

Fungal genomes are ten times bigger than bacterial genomes, and we need to read longer stretches of fungal DNA to stand any chance of identifying species. This is happening now with the aid of advances in DNA sequencing that allow faster and more accurate reads of longer strings of As, Ts, Gs, and Cs, along with the development of more sophisticated computer programs for analyzing the information gathered from fecal samples. Mycobiome research has also benefited from the efforts of investigators who have begun to discriminate between traces of fungi that are introduced with our food and the dominant species that actually run the active mycobiome (see the discussion of ghost gut fungi in the appendix).

Another obstacle to a more inclusive view of the gut microbiome is the relative scarcity of the fungi. With trillions of bacteria in the gut and *only* billions of fungi, fungi have been treated by bacteriologists as a minority group overseen by the ruling prokaryotes. This mathematical imbalance appears to discount the significance of the fungi in the chemistry of the gut until we factor in the relative bulk of the fungal cell. Revisiting the facts from chapter 1, the yeast cells that live in the gut are one hundred times bigger than the bacteria and present a huge collective surface area for interactions with the body. This more myco-centric view of the gut is changing the ecological description of the body and has significant implications for our health and well-being. As reliable information begins to emerge from mycobiome research, we are discovering that the fungi are game changers in gastroenterology.

GEOGRAPHY AND THE MYCOBIOME

A revealing study from China compared the gut mycobiomes of Hong Kong residents with people of different ethnicities from Yunnan.[1] Yunnan Province is in Southwest China and is home to twenty-five ethnic groups that speak multiple languages and consume a glorious range of traditional foods including bitter fruits and vegetables, flowers, pickles, yak jerky, and insects. Fungi are attached to all of these foods, and people in Yunnan add more fungi to their diet by cooking the wild mushrooms that grow in abundance in this region. Wild food consumption increases the diversity of the fungal DNA sequences detected in the fecal samples from Yunnan, but, crucially, this does not mean that more kinds of fungi actually live in the gut. The molecular techniques are so sensitive that we can amplify traces of rare organisms passing through the gut that have no influence on health at all. For this reason, it is essential to apply a data filter that excludes the weakest signals so that we can concentrate on the fungi that matter. When we do this, we find some interesting patterns among rural and urban populations in China.

An overwhelming difference between the gut fungi in Yunnan and Hong Kong is seen in the abundance of two species of yeasts: *Saccharomyces cerevisiae*, the food yeast, and a species of *Candida*.[2] The study shows plenty of *Saccharomyces* in the Hong Kong population and very little *Candida*; the reverse is true for the rural residents of Yunnan. The difference arises from life in the city, because we find that people from the various ethnic groups that migrate to Hong Kong lose their *Candida* and gain *Saccharomyces*. The plot deepens when we examine the health of the rural population. Analysis of blood samples reveals that the urban residents have better indications of liver function than the rural Chinese. This benefit seems to be related to the prevalence of the food yeast in the gut rather than abstinence from alcohol. The growth of *Candida* in the rural population seems to have some positive effects too, because more of this yeast was associated with higher levels of good cholesterol and lower levels of obesity. So, each fungus comes with its own benefits. The food yeast, *Saccharomyces*, in the Hong Kongers comes from baked goods and processed foods and may not be active as it passes through the gut, but this makes no difference in the mycobiome. Even if they are dead, the cells of fungi carried with food can change the chemistry of the gut if there are plenty of them, stimulate the growth of bacteria, and activate the immune system.[3]

Moving westward, Sardinia is designated as a Blue Zone country, where people enjoy unusually long and active lives. There are ten times more centenarians per capita on this Italian island than in the United States. Diet, genetics, daily exercise, and social cohesion are some of the presumed explanations for the longevity of these people, and indeed, the liveliness of the older Sardinians compares rather favorably with their contemporaries shuffling around the American Midwest. But what about their mycobiomes? *Saccharomyces* shows up in the Sardinians of all ages, along with *Penicillium*.[4] *Penicillium* is a distinctive member of the gut mycobiome that usually grows as branching filaments rather than budding yeast cells. It seems likely that this is another food import, because it is critical in cheese fermentation and Sardinians eat a lot of cheese made from sheep and goat milk. Many other fungi rise and fall alongside the species carried in the food, but it is not clear whether any of them are associated with longevity.

When we look at the fungi in fecal samples of Americans, we find the same fungi or close relatives of the Chinese and Sardinian species, with the addition of *Malassezia* yeast that we encountered as a skin resident in chapter 3.[5] Geographical differences in the gut fungi found in South Africans appear to be related to diet, but these are ancillary to the universal gut yeasts.[6] Taken together, these studies leave us with a picture of a core mycobiome within the gut that is assembled from a small number of fungal species that show up all over the world, whatever food we eat. Regional diets add or subtract from this microbial foundation, affecting the relative number of cells of different species, but *Candida* and a few of its compatriots are always there. The gut fungi seem to be very resilient and adapt to the continuous changes in the quantity and qualities of the food processed in the gut after each meal. These onboard yeasts act as buffers against the wholesale dissolution of the mycobiome even in cases of severe gastrointestinal illnesses. This reboot of the core mycobiome is accomplished by the rapid multiplication of the survivors after a population crash.

The discovery of a core mycobiome is very useful because this community of the commonest fungi found in the healthy gut can be compared with the mixtures of fungi that develop in various diseases. Recent research on the mycobiome has improved on the earlier surveys of the kinds of fungi found in the gut and is beginning to show which ones are present in the largest numbers and how they are interacting with the bacteria and the immune system. Through these studies, it has become clear that something is amiss with the mycology of the gut in many of the illnesses associated with diet and the function of the GI tract. A growing body of evidence suggests that fungi are involved in obesity, inflammatory bowel diseases, and even in the development of cancer. This has convinced some specialists that the fungi are a missing link in medicine.

OBESITY

This inquiry into the role of the gut fungi in disease begins with obesity, which affects more than one in ten adults, reduces mobility, and increases the risk of developing manifold illnesses. A lot of research on obesity involves fattening mice with a carbohydrate-rich diet. As the mice get heavier, they show metabolic changes, including fat buildup in the liver and accompanying alterations in the populations of fungi in their guts.[7] Even though some fungi increase and others decrease as the mice become fatter, the mycobiome does not wobble in a predictable fashion. The only thing we can say for certain is that the mycobiome is sensitive to rich diets and the resulting increase in weight. Even without any fattening, mice fed an "exotic" fungus that grows in yogurt produces an immediate disturbance to the resident fungi and bacteria in the gut.[8] These experiments highlight the sensitivity of the gut mycobiome to diet, with changes to the core communities of microbes resulting from a short-term or continuing change in diet or even from a single unusual meal.

Studies on human obesity have revealed some interesting changes in our onboard fungi, although there is nothing in these findings that offers any immediate help in the search for new approaches to weight loss. Although the fungal DNA from fecal samples indicates that there are different mixtures of yeasts in obese and non-obese people, these are quite modest. What we hope to find in a comparison of this kind is something that distinguishes the two groups of participants—a red flag of a microbe that shouts Weight-Loss Fungus! This germ does not appear to exist, although a study from Spain found that species of *Mucor*, the bread mold, were present in non-obese participants and rare in the obese cohort.[9] They went on to show that the absent molds began to appear in the obese participants when they followed a weight-loss diet and that the signals from a variety of *Mucor* species became more frequent as the patients lost weight. The activities of these fungi in the gut are unknown, but their responsiveness to the metabolism of the host parallels the reactions of other kinds of fungi in the mouse mycobiome studies.

The ebb and flow of *Mucor* and other fungi in the mycobiome as we gain or lose weight could be passive fluctuations in which the fungi play no active role. *Mucor*, for example, may be a commensal fungus, meaning that it lives in the guts of lean individuals without helping or harming them. It is also possible that some of the fungi reinforce the status quo by releasing chemical compounds that assist in the digestion of food in a way that serves as a buffer against weight gain. Other fungi may work to bolster the obese state, making it more difficult to lose weight even when the number of calories is reduced in a new diet. We need to know a lot more about the chemical communications between the fungi and the body before we can take this idea of a supervisory relationship for the gut mycobiome any further, but this is a distinct possibility.

GASTROINTESTINAL DISEASE MYCOLOGY

The handful of species of *Candida* that serve as the core of the gut mycobiome are part of the standard equipment of the human body—true symbionts, rather than transients, and we would probably be in trouble without them.[10] I say probably, because we do not know what would happen if these fungi were completely eliminated from the digestive system. *Candida* yeasts are a constant throughout life. Other fungi come and go as we get older and change our diets, but *Candida* is with us for the long haul. Problems develop when the number of these yeasts snowballs, which is what happens in inflammatory bowel disease (IBD).[11] IBD is an umbrella term for chronic inflammation of the bowel that can be manifested as Crohn's disease in the small intestine and ulcerative colitis in the colon and rectum.

The symptoms of IBD and the growth of *Candida* flare up in tandem, suggesting a causal relationship in which the illness simulates the fungus, and the fungus worsens the disease symptoms. This model of IBD is exciting from a therapeutic point of view. In a more straightforward fungal infection, we try to eliminate the fungus to resolve the tissue damage. In a chronic condition like IBD, the fungus is part of the healthy body, and the symptoms of the illness might be alleviated by repressing the fungus rather than ridding it entirely. This gentler approach to treatment could be achieved, at least in principle, by modifying the diet—if we could agree on the foods that pacify *Candida*, or by finding an effective supplement. We will look at dietary intervention at the end of this chapter, but there is another more direct strategy for revamping the fungi in the gut: fecal transplantation.

Fecal transplantation is a controversial treatment for digestive disorders in which small samples of feces (equal to three tablespoons) from healthy donors are implanted in the intestine of sedated recipients using a colonoscope. In an Australian study of patients with colitis, the transplants by colonoscopy were followed by repeated transfers by enemas administered by the patients themselves for eight weeks.[12] This intensive therapy produced remarkable results, with a decrease in ulceration in one-third of the patients and complete remission from their IBD symptoms. The strongest signal from the mycobiome was a decrease in the levels of *Candida* after fecal transplantation, and this was accompanied by an increase in the diversity of the bacteria. The mycobiome and the bacteriome—the fungi and the bacteria in the microbiome—were exchanging their footings in the gut. The greatest success was enjoyed by patients who showed the highest levels of *Candida* relative to other fungi before treatment. One interpretation of this finding is that there was so much of the yeast in the IBD patients that it was crowding out the healthy bacteria. Fecal transplantation seems to work by resetting the populations of microbes, which results in a decrease in inflammation that allows the intestine to heal. We certainly need more research on this surprising therapy.

Inflammation is the first line of defense against harmful germs, and we would be ruined without it. But when inflammation comes on too strong, the body is overwhelmed by the mobilization of immune cells and release of irritating chemicals, swelling of blood vessels, sensation of heat, and other symptoms. Too much stimulation can be lethal. Too little is deadly too. This is the Goldilocks principle of immunology. In a healthy digestive system, the immune system is responsive to any signs that fungi and bacteria are infiltrating the tissue of the gut wall, and our cells are probably patching and repairing tiny nicks and leaks all the time. This tempo is just right. Colitis and other illnesses develop when the inflammation spreads over larger areas of the gut wall and the tissue damage overwhelms the repair mechanisms. Continuous or chronic inflammation is a disastrous instance of unconscious self-injury. We should think about this when we see advertisements about health foods or herbal medicines that promote optimal health by stimulating the immune system.[13]

The immune system is alerted to the presence of fungi by the distinctive nature of the molecules on their cell surface. These trademark compounds include enzymes and other proteins decorated with sugars that we call mannoproteins. The immune system is programmed to react to these flags from the time we are born and the mycobiome begins to develop on the skin and inside the digestive system. One of the puzzles about the mycobiome is why the fungi that start growing on the body in infancy begin to damage the gut wall in some adults. The answer lies in a combination of factors, including other health conditions that may weaken the tissue barriers and allow the fungi to overcome the normal obstacles to infection. This vulnerability may have a genetic foundation and increase with aging or it may follow a viral infection. A very provocative study from China revealed a surge in the number of *Candida* cells in the guts of patients hospitalized with COVID-19 or with influenza.[14] These changes to the mycobiome have the potential to upset the immune system, trigger inflammation, and even increase the likelihood of fungal infection. It sems that viruses can impact the ecology of the whole body by agitating the fungi and bacteria in the gut.

Fungal infections can spread from the digestive system to other parts of the body when the immune system is damaged, and *Candida* reveals itself as a lethal enemy in these cases of disseminated disease. *Candida* does this by switching from growth as a budding yeast to extending as filamentous hyphae that pierce the gut wall and reach the interior of blood vessels. The fungus switches from yeasts to filaments to move beyond the bowel, back into yeasts to tumble in the bloodstream, and then again into invasive filaments when it is delivered to solid tissues by the capillaries. This flipping between growth forms occurs whenever *Candida* escapes from its settled life as a benign resident in the body, where it is kept in check by other fungi, its bacterial neighbors, and the immune system, and engages in a population explosion. We are the unfortunate hosts for this ecological exercise. The temperamental behavior of the fungi is as clear in the gut as we have seen elsewhere in the body.[15]

The disruption of the mycobiome in irritable bowel syndrome (IBS) is very similar to the dysbiosis measured in IBD, with an overall drop in the diversity of the fungi and an increase in *Candida*. A study on fungi and IBS from the Netherlands also found an increase in *Saccharomyces*.[16] IBS is usually treated as a less serious complaint than IBD because it is not accompanied with continuous intestinal inflammation. Even so, the symptoms of these illnesses overlap, and IBS can be debilitating. Some doctors believe that IBD and IBS are manifestations of the same illness with different degrees of inflammation. Anxiety and depression are described as comorbidities for both conditions, meaning that patients diagnosed with IBD and IBS are more likely to suffer from these mood disorders. This is reminiscent of the claims about asthma and anxiety discussed in chapter 3 and the impossibility of separating cause from consequence.

FUNGI AND CANCER

Inflammation is a big player in the development of tumors, which raises the possibility that the mycobiome may be involved in some cancers.[17] Early diagnosis of IBD, before age thirty, is a risk factor for colorectal cancer, suggesting a progression from severe inflammation of the gut wall to the formation of noncancerous polyps called adenomas, and the later transformation of these growths into malignant tumors. The mycobiome changes as the health of the gut declines, with shifting ratios of different groups of fungi, but we do not see a universal pattern. One study showed a rise in the level of *Malassezia* yeasts, rather than *Candida*, and analysis of polyps revealed other fungi related to *Malassezia* that tend to be better represented on the skin. But the absence of a common denominator among the revolving populations of fungi indicates that the gut fungi are responding to inflammation in IBD rather than causing the illness. This does not mean that the fungi are irrelevant—far from it, because the chaotic mycobiome may exacerbate the tissue damage resulting from the tumors developing in the wall of the intestine.

The active participation of the mycobiome in the progression of cancer is supported by recent studies in which fungi have been detected along with bacteria inside tumors in the intestine, pancreas, lungs, and other body sites.[18] DNA from *Candida* and *Saccharomyces* yeasts is found in the tumors in all of these tissues, with *Candida* predominating in most places. We know that this DNA has not strayed from outside the tumors because microscopic images show fungal cells inside cancer cells and macrophages. Live cells of the fungi have even been isolated from tumors of the colon and grown in culture dishes. These findings show that there is nothing passive about the fungi within tumors. They are growing within the cancerous tissues, insinuating themselves between the cancer cells, and some are being consumed by the macrophages of the immune system. Fungal DNA is also found in the bloodstream of patients with late-stage metastatic cancer. This suggests that the fungi are leaking through the weakened capillary walls that also allow the cancer cells to migrate to other parts of the body.

The investigators involved in this research believe that the destabilization of the mycobiome and the entry of fungal cells into tissues can be used as new markers for the development of cancer. This means that the identification of fungal DNA in a blood test or a biopsy sample could be useful in determining the developmental stage of a tumor and helping patients and doctors decide how to proceed with treatment. Whether the fungi are misbehaving in tissues damaged by the growth of a tumor (consequence) or stimulating its development in some way (cause) remains to be seen.

MYCOLOGICAL MIGRATIONS:
ORAL AND GENITAL FUNGI

Although different mixtures of fungi are found in the gut mycobiome and the skin mycobiome, yeasts spread from one to the other at the beginning and end of the digestive system. Skin fungi settle into the mouth and gut fungi seep into the vagina. We begin with the mouth. A peck on the cheek transfers fungi from person to person, and more intimate contact via lips and tongue adds to the oral broth. Spores are vacuumed from the air as we breathe, and the mouth is smeared with fungi clinging to food and floating in drinks. Few of these nomads survive, but the strongest yeasts join the populations swarming along the gumline and coating the tongue and palate. A founding dollop of a few hundred yeast cells swells into the millions in a few hours if the conditions are perfect, but the mouth is a demanding place. Beyond the attentions of the immune defenses, fungi must cope with the climatic swings of the mouth as we clear our throats and speak, flood the oral cavity with hot and cold liquids, and attack the biofilms with a toothbrush, until, at last, we leave the mycobiome alone and embrace the sandman. As we snooze, the fungi are at work, feeding and reproducing, poisoning some bacteria, cooperating with others, and preparing for sunrise and the first cup of coffee.

Preparation on the part of a fungus sounds fantastical, but the sensitivity of fungi to their surroundings is well established, and experiments show that they possess the rudiments of memory that enable them to ready themselves for a stressful change in conditions.[19] This behavioral complexity is demonstrated by the fungal response to salt exposure. When the salt concentration in a culture is increased, yeasts respond with a flurry of biochemical changes to resist dehydration, and the mistreated cells respond more swiftly when they are aggravated a second time. We say that they are primed, and this protects the cells against other harmful treatments including heat shocks and flushing with hydrogen peroxide (which acts as a damaging oxidizing agent). It follows that this simple response could allow the fungi of the oral mycobiome to buffer themselves against the morning espresso bath. Coffee is a problem for fungi because it damages their DNA, and some yeasts have a transport system that expels caffeine if it gets into the cell. By anticipating the morning brew, these fungi have an opportunity to fortify their membranes before the sun rises to keep the bitterness at bay. Fungal sensitivity and consciousness are revisited in chapter 10.

Most of the fungi that live in the mouth are the familiar residents of the mycobiome found in other locations. Yeasts are prominent, and *Candida* and *Malassezia* dominate a pair of microbial communities that we call mycotypes.[20] When *Malassezia* is the most abundant fungus, it is associated with a rich mixture of bacteria and other fungi. This mycotype is found in healthy mouths. In smokers with active tooth decay and other indications of poor oral hygiene, the *Candida* ecotype emerges, in which the diversity of other fungi and bacteria falls. The healthy *Malassezia* mycotype is replaced by the unhealthy *Candida* mycotype.[21] The rise of the second population of yeasts is a reliable sign of dental problems and is found in children with tooth decay and in older people who have lost their teeth and wear dentures. Rather than causing tooth loss, *Candida* may reinforce a wider decline in oral health by reducing the growth of protective fungi and bacteria and resisting natural repair mechanisms.

The problems multiply with the formation of resistant coatings or biofilms of bacteria and fungi on the tooth surface. These layers of microorganisms thicken and become acidic when sugars stimulate the growth of *Candida* that nestles alongside the bacteria. Biofilms are involved in tooth decay, and the stimulatory effect of sugars may explain why children who overload with candy are prone to cavities. Studies on elderly patients also support the link between diet and changes in the oral mycobiome that may accelerate tooth decay: much higher levels of *Candida* were found in saliva from patients in Japan and the Netherlands who wore dentures compared with their peers with natural teeth.[22] Interestingly, levels of *Candida* also tend to be higher in older people who live in nursing homes compared with those who live in their own homes. Diet may be a less significant factor here, with the presence of other health problems in people who move into nursing homes having a knock-on effect on the oral mycobiome.

Moving to the vagina, *Candida* is the biggest player in this ecosystem. The commonest species is *Candida albicans*, which is detected in at least 20 percent of women and is responsible for vulvovaginal candidiasis, with symptoms of itching, burning, and discharge. Some antibiotics stimulate the overgrowth of this yeast by eliminating the bacteria that help manage the fungi in the normal community of microbes on the body. Recurrent candidiasis, which is defined as four or more episodes per year, affects hundreds of millions of women.[23] This can be a debilitating condition. There is no cure, and treatment relies on suppressing the growth of the fungus with antifungal medicines. Annual medical spending on this complaint soars into billions of dollars in the United States alone. In the limited studies of the vaginal mycobiome, all of the fungi identified on this part of the body surface are species of *Candida*. Healthy populations of *Candida* have a protective effect against the development of intrauterine adhesions by modulating the growth of bacteria and less common fungi in the cervical canal and middle vagina that can damage these sensitive tissues.[24] Men are also affected by this yeast, albeit rarely, when candidiasis develops on the surface of the penis following contact with a partner with the vaginal form.

When we consider the importance of *Candida* in the oral mycobiome, its dominance of the vagina, prevalence in the gut, and presence on the skin, this yeast materializes as the most important fungus in human health. We have seen the power of many other fungi in supporting and collapsing our health, but in terms of the overall operation of the human ecosystem, this yeast is unchallenged as the fungus at the heart of the human being. We are human yeasts, *Homo mycosapiens* or *Homo fermentalis*.

RIPPLE EFFECTS OF THE GUT FUNGI

With mycobiome research in the early stages of sorting fact from fiction, or facts with an impact on health from those without, the case for a link between gut fungi and illnesses whose symptoms are expressed in other parts of the body is building. Some of these claims seem unlikely at first, but the revelations about the fungi that live in our bodies are so surprising that we should keep an open mind. If my mycology professor had told me when I first looked at fungal spores through a microscope in the 1980s that some of these tiny grains belonged to fungi that could be found in the colon, I would have thought him batty.

Vaginal delivery and breastfeeding have a protective effect against the development of asthma in children and are associated with different gut microbes than the ones found in C-sectioned and bottle-fed babies (see chapter 1). Problems in the development of the immune system in asthmatic children offer a hypothetical link to the gut microbiome and shifts in the numbers and kinds of yeasts that are seen in asthmatic children.[25] Early treatment with antibiotics is another risk factor for asthma, which strengthens the fungal connection according to the following chain of logic: fungi can be highly allergenic → fungi grow in the infant gut → gut fungi are disturbed by antibiotics → infants that receive antibiotics are at greater risk of developing asthma ⇒ [*ergo*] disturbing the mycobiome can lead to asthma.

An obvious obstacle to embracing the idea that fungal unrest or dysbiosis in the gut can provoke asthma is the physical separation of the digestive and respiratory systems. Eating does not affect breathing in any noticeable way, and breathing dust does not cause intestinal distress. This segregation of organs is an essential element of the way that medical students learn about human anatomy, lungs this week, kidneys next, although the connections should be emphasized as often as possible. For example, the immunological systems that protect the gut and the lung are coupled through the blood and lymph vessels, and we recognize this intimacy with the term gut-lung axis.[26]

Similar ripple effects between the gut mycobiome and illnesses presented in other parts of the body have been proposed for other conditions that seem unrelated to gut function. The number of budding *Candida* cells soars in type 1 and type 2 diabetes, and multiple species of *Candida* appear in the guts of children with type 1 insulin-dependent diabetes. The overall diversity of fungi in the gut mycobiome also seems to increase in some liver diseases and decrease in others. *Candida* flares are seen in hepatitis, cirrhosis of the liver, and a disease called primary sclerosing cholangitis (PSC) that damages the bile duct.[27] PSC is also known as Walter Payton's disease, after the famous American football player who died in 1999 from bile duct cancer caused by this condition. Multiple sclerosis is the most controversial of these putative correlations, with different patterns of fungal abundance seen in patients with this chronic autoimmune disease.[28] In the absence of a consensus in the mycobiome experiments, it seems most likely that the fungi are responding to the gut inflammation that develops in multiple sclerosis patients rather than playing any causal role.

OPTIMIZING THE GUT MYCOBIOME

There is a tendency to view the body as a temple ruined by modernity. This viewpoint is entirely reasonable when we consider the frightful nature of processed foods saturated with sugar and salt, lacking all traces of fiber, and delivering the calories required for weight maintenance by a sasquatch rather than a slothful human. Those with sufficient tenacity and affluence to eat a more balanced diet that emphasizes plants and get some exercise can feel happier in the twenty-first century and reflect on the likelihood that they are likely to live longer than most representatives of our species throughout history.

So, what can we do to promote a balanced mycobiome in the gut that may support our overall health? The fungible concept of the healthy mycobiome makes me wary of any dietary recommendations, and my fungible waistline is unlikely to inspire confidence in my readers. Other commentators on the mycobiome are bullish enough to recommend food and behavioral choices to improve the mycobiome, attain digestive wellness, and lose weight. Mahmoud Ghannoum, a distinguished medical mycologist at the University Hospitals Cleveland Medical Center, is the most prominent advocate of mycobiome manipulation, which he explains in his book *Total Gut Balance: Fix Your Mycobiome Fast for Complete Digestive Wellness* (2019).[29] Eve Adamson, an award-winning author on dieting, is credited as a cowriter, and the book includes sixty "Mycobiome Diet" recipes. There is a lot of useful information on fungi in *Total Gut Balance*, and Ghannoum explains that *Candida* is a harmless resident until dysbiosis sets in and the fungus reveals its darker side. The science is leavened with advice about smoking cessation, yoga, and meditation, which is very reasonable but strays from the mycobiome. The recipes look very appetizing in the photographs, but the only study on the impact of the Mycobiome Diet on the mycobiome was published by Ghannoum himself in the *Journal of Probiotics and Health*.[30]

Pending advances in mycobiome research, there is little dietary advice supported by scientific evidence that will guarantee a healthy mycobiome. Healthy people tend to have healthy mycobiomes, and a great range of illnesses are accompanied by unhealthy mixtures of fungi and bacteria. There is a frustrating circularity to this relationship and few ways to intervene and use the onboard microorganisms to treat disease. *Candida* is the obviously problematic fungus in many debilitating conditions that involve the mycobiome, and dampening this yeast when it grows to the exclusion of everything else could be helpful. The effective treatments for vaginal yeast contain low doses of antifungal drugs and are applied as topical agents. Use of the same medicines to treat *Candida* in the gut has the unwelcome effect of damaging the other fungi whose growth we want to support. We do not want to eliminate any organism that might be an essential partner in the healthy human ecosystem. Coconut oil, undecylenic acid (from castor oil), and oregano leaf extract are dietary supplements that are advertised as treatments for yeast overgrowth. Coconut oil, which is recommended in the Mycobiome Diet, has a proven effect on the mycobiome of mice, reducing *Candida* overgrowth and restoring the healthy mixture of fungal species.[31] This is not much to go on, but with few side effects, coconut oil is worth trying as a dietary remedy. Reducing *Candida* through dietary changes seems unlikely to resolve any serious condition like IBD, but any relief from the crippling symptoms would be marvelous.

The key to a healthy mycobiome may lie in preventing an unhealthy one from developing in the first place, and the most direct way to satisfy the needs of our fungi is to feed them properly. While the Mycobiome Diet totters on a weak scientific foundation, the recipes and other advice offered by Mahmoud Ghannoum push readers in the right direction, away from processed foods and toward a sensible diet that tastes good and emphasizes plants. If your digestive system operates like a well-oiled machine, pay attention to what you are eating and keep doing so. If it misbehaves, interrogate someone with a happier gut and find out what they have been eating. Clearer advice about optimizing the mycobiome will come as this scientific inquiry moves on from naming the fungi in the gut to figuring out what they are doing in there. This will be expensive and time-consuming and require collaborations among specialists in gastroenterology who know the intestine, bioinformatics experts who read DNA for a living, immunologists who understand immunology, and mycologists infatuated with fungi.

From this inward examination, in part I, of the fungi that live on the body for better or worse, supporting and damaging our health, we move outward, in part II, looking at the ways that we interact with fungi outside the body. We begin with chapter 6, on mushrooms, molds, and yeasts in the diet.

PART II

Outward

6

Nourishing

MOLDS AND MUSHROOMS IN OUR DIETS

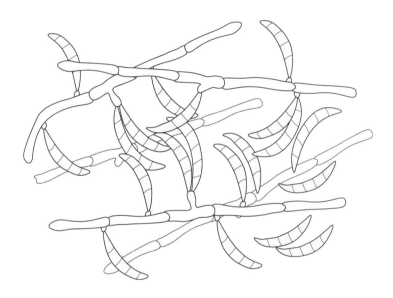

MILK CURDS are stiffened and blue-veined in the coolness of caves; sausages are bloomed with white powder as they hang drying on strings; beans and cereals dissolve into soy sauce and jellify into tempeh; bread dough rises; and grains and grapes are transformed into beer and wine. Humans crafted these foods for millennia without any idea that microscopic fungi were threading through cheese and bubbling in vats. All they knew was that their diets were invigorated by experimenting with raw foods, and they marveled at the handiwork of their gods. Biotechnology began as this monkish pursuit. In our time we are applying the same genetic techniques that have identified the fungi that grow on the body to understand the ecology of food. Who would have guessed that a cheese wedge is one of the wonders of the microbial world? In this chapter we explore the foodie extension of the human-fungus symbiosis that takes us from the body to the farm and to fermentation towers brimming with mycelia that make chicken nuggets without chickens.

Penicillium is the cheesemaker. The name of this mold, which means brush, refers to its bristly stalks topped with chains of spores that resemble tiny dreadlocks.[1] In nature, these spores are blown into the air or catch on the hairs of passing insects, and each particle of fungus carries the genes for making a mycelium in a new location. Most spores land in places without food and water, where they shrivel and die, but a small proportion survive and go on to craft the next generation of spores. And so it goes, and has gone on, from spore to spore for millions of years, conveying the instructions for making this fungus.[2] Cheesemakers circumvent the wild dispersal mechanism and add the spores of the fungus directly to their milk curds.

Penicillium is the first Latin name or genus of hundreds of fungal species that grow everywhere and feed on everything, spoil food, make toxins and antibiotics, and flavor cheeses and preserved meats. Another genus, *Aspergillus*, is equally influential on agriculture, medicine, and food. *Penicillium* and *Aspergillus* are filamentous fungi that do not produce mushrooms or any macroscopic fruit bodies at all. They are the iconic molds, or "moulds," in British spelling. Along with the brewing and baking yeast *Saccharomyces cerevisiae*, these microbes occupy the top spots in the human-fungus symbiosis: not in terms of the mycobiome and human health—*Candida* takes that award—but as prizewinners for supporting civilization.

BREWING AND CHEESEMAKING: THE FIRST FERMENTATIONS

To understand the best-partner nomination for *Penicillium*, we begin with a visit to a salt mine in Austria and look at the feces left by an Iron Age miner more than 2,600 years ago. This archaeological treasure was left in a mineshaft in the mountains that tower above the beautiful alpine village of Hallstatt. People were digging here for thousands of years before the miner took his historic bathroom break, and this amazing mine remains active today. The sample of paleofeces is the size of a grape, and it was preserved by the saltiness of the mine. It is very fibrous in structure, which is consistent with a high-fiber diet that included barley and other cultivated cereals whose remnants are visible under the microscope. The miner had also eaten apples paired with Roquefort cheese and quenched his thirst with beer. He was a proto-gourmand whose repast would not be amiss in a fashionable bistro today. There are no physical traces of the cheese and beer, of course, but analysis of DNA in the paleofeces by a team of researchers led by Frank Maixner at the Institute for Mummy Studies in Italy identified the cheesemaking fungus, *Penicillium roqueforti*, and brewer's yeast, *Saccharomyces cerevisiae*.[3] Both fungi had become part of the salt miner's gut mycobiome, in the same way that every fungus that we eat flows through the digestive system, interacting with the intestinal bacteria, triggering immune reactions, and affecting our health. Maixner's study provides a snapshot of the unfolding relationship between humans and fungi in Europe before the Roman conquest—the way that the diets of laborers were enriched by fungal fermentation.

Our collaboration with fungi began with the transformation of sugars into alcohol by yeast, symbolized as $C_6H_{12}O_2 \rightarrow CH_3CH_2OH$, which is the most consequential chemical reaction in human history.[4] My hunch is that palm wine was the archetypal brew in tropical Africa, where people found that the sweet sap drained from palm trees fermented spontaneously in the sun. This idea is supported by the discovery of fractured starch grains from wine palms and sorghum on the surface of 105,000-year-old stone grinding tools in a cave in Mozambique. It is tempting to think that the early humans were processing palm stems and grass seeds to make the kinds of artisanal brews that remain popular in Africa today.

The oldest clear evidence of brewing comes from the analysis of pottery remains showing that rice wine was fermented in Chinese villages nine thousand years ago.[5] Wherever it began, brewing was the first deliberate application of the chemical wizardry of the fungi. This was deliberate in the sense that the brewers learned to prepare their ingredients in a way that stimulated fermentation, although they obviously had no concept of the invisible yeasts and molds that performed this alchemy. Indeed, early brewers interpreted the formation of alcohol through their religions and showed their appreciation by worshipping festive deities that included the Daoist god Yidi in China, the Sumerian beer goddess Ninkasi, and the Yoruba spirit Ogoun in Africa.

The extension of human-fungus symbiosis from brewing with yeasts to cheesemaking with molds was the next technological breakthrough that bears the stamp of serendipity. Cattle and other mammals were originally domesticated to furnish settlements with a reliable supply of meat. Milking was a later inspiration. Milk is rich in fat and protein, but genetic research shows that the early herders belonged to lactose intolerant populations and would have found fresh milk unpalatable.[6] Butter was one of the early solutions to the lactose problem, which concentrates the milk fat and allows a good deal of the lactose sugar to drain away in the buttermilk. Cheesemaking works in a different way, by separating the fats in the curds and expressing a lot of the sugar in the whey. By churning butter and fermenting cheeses, herders produced palatable and nutritious foods that took up less space than gallons of milk and lasted a lot longer before spoiling.

Yogurt was another early dairy product that is fermented by microbes. Residues in Neolithic pottery show that yogurt drinks were fermented from mare's milk in Central Asia and that cheeses were made from cow's milk in northern Europe at least seven thousand years ago. In his *Histories*, written around 430 BC, Herodotus described the Scythian use of mare's milk to make *kumis*, which is a mildly alcoholic drink that remains popular in Russia and other countries. Lactic acid bacteria dominate yogurts, and yeasts are unwelcome because they spoil the taste. Yeasts are more amiable in *kumis*, where they produce alcohol and create the frothiness of this drink, which has been called the champagne of the steppes.

BLUE CHEESES

By the time the armies of Julius Caesar conquered Gaul in 50 BC, cheese had become a widespread part of the Western European diet. In the first century AD, Pliny described a range of cheeses imported to Rome from its empire in his *Naturalis Historia,* and Petronius detailed a soft wine-soaked cheese served at the end of a banquet in his *Satyricon.*[7] Pliny praised the cheeses from southern France without specifying Roquefort (as some promoters have claimed), but blue cheeses were produced long before then, as the salt miner testified through his ancient deposition.

According to legend, Roquefort cheese was discovered by a French shepherd who put his bread and cheese in a cave for safekeeping when he ran off in pursuit of a young maiden. Months later, he returned to the cave and found his cheese metamorphosed into a blue-marbled delicacy. There have been, no doubt, many distracted shepherds, but science has shown that the Roquefort legend is a crock. Contrary to long-standing ideas about this *Penicillium* evolving as a cave dweller, it is a domesticated variety of a fungus that rots vegetation and thrives on farm silage stored as winter feed for livestock.[8] The only reason that it grows in caves is because its spores are taken into them by cheesemakers in the commune of Roquefort-sur-Soulzon in southern France and added to curds clotted from ewes' milk. Here, the curds are shaped into wheels and allowed to drain before salting and spiking with needles to aerate the cheese. This stimulates the growth of the mold, along with complementary bacteria. Afterward, the cheeses mature in caves for at least three months as the fungus performs its sorcery in the cool humid air. As the wheels age, their veins are colored by the blue-green spores of the fungus.

By transferring cultures of the fungi that produced the most successful batches of Roquefort onto chunks of moldy bread, early cheesemakers selected a strain of *Penicillium* with the strongest combination of characteristics. They wanted one that grew slowly to ensure a steady maturation process and produced plenty of spores that would serve as a vigorous starter for fresh batches of cheese. These artisans were engaged in the practice of artificial selection that we usually associate with the domestication of plants and animals. And as the fungus got better at making the blue-veined cheese in the caves, our amorous shepherd was contributing to the Roquefort story by choosing lambs from ewes of the Lacaune sheep with the richest milk and puppies from sheepdogs that were the best at herding them. Across the centuries, selective breeding worked its magic on the molds, sheep, and dogs to craft the perfect cheese.

Penicillium and other fungi work in combination with bacteria to create masterpieces like Roquefort. Bacteria break down the lactose in milk, and the accompanying acidification promotes coagulation. Fungi added to the resulting curds feed on the milk fat, which controls the consistency of the cheese, and they are also responsible for producing the distinctive aroma and taste during the maturation process. The strain of the fungus that makes Roquefort in the eponymous French township is maintained in laboratories today to ensure its unwavering behavior when it is added to the curds. Other strains of *Penicillium roqueforti* create the sweet, nutty crumbliness of Stilton from England and subtler creaminess of Danish Blue from cow's milk. Cashel Blue from Ireland is another win for this species of fungus, but milder blue cheeses, including Gorgonzola and Blue d'Auvergne are made by a relation called *Penicillium glaucum*.

Another mold, *Penicillium camemberti*, makes the soft white crusts of Camembert and Brie.[9] These creamy cheeses originated in northern France. Brie was produced in the Middle Ages, and its crust had a blue-gray color until the twentieth century, before cheesemakers selected for strains that form a pure white surface. The growth of the mold is controlled by refrigeration, and the crusts are flattened by their wrappers, which stifle the slow expansion of the fungus. When the wrapper is removed, its folds are left as impressions in the white fungus that has grown along the seams, seeking opportunities to escape. If the unwrapped cheese is left at room temperature, it can turn fluffy. This is nurtured in some cheese varieties, like the firmer and hazel-nutty Saint-Nectaire, which becomes covered with *poil de chat*, or cat's hair. The fungi that grow the hairs and darken the cheese surface are species of *Mucor*, extending the activities of these versatile species of fungi from the gut mycobiome, where they are common occupants, to the microbial communities of the cheese and back again when we eat them.

Saint-Nectaire, which comes from Auvergne in the middle of France, is fermented and ripened by an exceptionally complicated succession of fungi and accompanying bacteria.[10] Multiple species of yeasts start by growing in the milk curds, breaking down the milk fat and reducing the acidity, which allows bacteria to multiply; molds join the fray as latecomers to firm and fluff the crust. Ten billion microbes can grow in a gram of cheese.[11] The populations of different fungi and bacteria rise and fall as the cheese ages, reflecting changes in the availability of nutrients and the give and take between microbes that ranges from cooperation to chemical warfare. The intricacy of some of these interactions is astonishing. One mind-bending study on Saint-Nectaire shows that bacteria use the fungal hyphae as physical guides, like miniature railroad lines, for high-speed travel through the rind.[12] This collaboration is limited to certain types of bacteria, which has the effect of controlling the mixture of microbes in the maturing cheese. Other microbial performances, most of them unknown, animate—or, to be more precise, *fungate*—the taste, smell, color, and texture of every cheese, whose brilliance arises from their dalliance with death, edging so close to decomposition yet brimming with life. A cheese is, as American author Clifton Fadiman wrote, "milk's leap toward immortality."[13]

The removal of lactose and production of a compact and spoilage-resistant food explains why cheesemaking began in the Neolithic. The availability of this high-calorie dairy product must have been a godsend during the winter months in northern latitudes when fresh meat and vegetables were scarce. Cheese has another great advantage over raw milk in its safety. The *Listeria* bacterium that grows in raw milk is inhibited by the activity of the preferred bacteria and yeasts in cheeses. Artisan cheeses made from unpasteurized milk are not immune from this dangerous form of spoilage, but cheesemaking has become such an exacting business that cases of listeriosis from cheese are extremely rare. If we accept the minuscule risk of listeriosis from these raw milk products, we access new dimensions in the cheese universe, hundreds of sensational varieties that may have the additional benefit of contributing to a healthy digestive system when samples of cheese microbes merge with the gut microbiome.[14]

None of these prehistoric advantages of cheese explain why we are attracted to the strong flavors of blue cheese. To understand the pleasures of Roquefort and its rivals, we need to think about the evolution of our sense of smell. The fungi in blue cheeses produce volatile compounds that belong to the groups of chemicals that perfumiers mix in their fragrances. We find these alluring for the same reasons that we are attracted to zesty citrusy odors, whose underlying appeal evolved as a stimulus for finding wild fruits. Fungi are not interested in attracting us with these smells and seem to make these volatile chemicals to repel other microorganisms.[15] The explanation for turophilia (cheese-loving) becomes more complicated when we consider the popularity of washed-rind cheeses, whose aromas are produced by bacteria. Consider the shockingly sudoriferous Limburger from Belgium, whose smell is produced by a bacterium that is a culprit in foot odor. The taste for Limburger and other maximally pungent cheeses, like Stinking Bishop and Époisses, probably lies somewhere in our sexual attraction to certain body odors and pheromones. The bewitching power of this mephitic matter on other animals is illustrated by the effectiveness of Limburger as bait for trapping malarial mosquitoes that are programmed to respond to the same signals in human sweat.[16]

FERMENTED MEAT, FISH, CEREALS, AND BEANS

Moving from la fromagerie to la charcuterie, we find more evidence of the culinary skills of *Penicillium* in the whitish bloom on the skins of the dry-cured meats hanging from strings. Some meat producers inoculate their sausages by dipping them in a starter culture of spores before they are placed in drying rooms to mature. Others allow their sausages to become colonized by fungi without any human interference. Either way, the fungi change the taste of the sausages as they feed on the fats and proteins in the meat and contribute to the characteristic aromas of these regional delicacies. The fungi also absorb water from the raw meat, which accelerates the drying process. They also fend off other molds that cause spoilage, including species that produce toxins. *Penicillium nalgiovense* is the commonest fungus used to cure meats and is used by the major manufacturers.[17] This fungus is joined by *Penicillium salamii* on Italian salami, soppressata, and capocollo, whose spores float around in the rooms where the raw meat is packed into casings.[18] The same fungus is obviously an expert in meat curing because it also grows on air-dried ham in Slovenia.

Because fungi and their spores are everywhere on the lookout for food, and because they are so skilled in decomposing the rest of nature, they colonize every kind of raw food that we harvest and compete with us for all of the available calories. The resulting spoilage is terrifically harmful from the perspective of the farmer and consumer, but this is balanced by the protection afforded by the elite group of food fungi whose growth we have encouraged for thousands of years. All of the uses of fungi to protect and modify raw foods must have been discovered by accident and perfected by experimentation. This is exactly what happened with brewing and cheesemaking, where nothing is left to chance today.

Regional loyalties to some fermentations are more difficult to comprehend than others. Fungi and bacteria that ferment Greenland shark meat make it safe to eat without making it appealing to eat. The resulting chewy snack, which is an Icelandic treat called *hákarl*, has a memorable bouquet of stale urine. It is one of the wonders of the culinary world, rejected by the palates of most tourists who try a cube of the stuff during a visit to the volcanic island. The preparation of *hákarl* has not changed in any fundamental way since the Viking Age because it is crafted automatically by microbes that are ready to pounce on the flesh as soon as it is exposed to the air. Icelanders ferment the shark flesh in tubs with perforated bottoms that allow the fluid to drain away for up to six weeks. After this preparatory relaxation, *hákarl* is left dangling in drying sheds for a few months before it is eaten. The raw shark meat is poisonous and is detoxified during this lengthy polishing, which is an undeniably good thing. When a sample of the final product is spiked with a toothpick, it might be mistaken for a piece of cheese until one is distracted by the sense that somewhere close by, a cat with a bladder infection has relieved itself with great enthusiasm. This impression of feline incontinence is replaced by much ghastlier scenarios if you proceed to put it in your mouth. Hoping that this book will sell well in Iceland, I should add that these comments are based on online reviews rather than personal experience. Every country has its culinary quirks, and I am sure that plenty of Icelanders who enjoy *hákarl* are revolted by Marmite, the salty black paste that is revered in my birthplace. Marmite also has its place in this chapter, as a fungal product that is formulated from yeast cells exhausted from the labor of crafting beer.

Hákarl is fermented by a chaotic mixture of bacteria, yeasts, and molds, and this ensemble of microorganisms—the *hákarl* microbiome—differs in every sample of the snack sold in stores.[19] This is a change from the microbiomes of blue cheeses, in which one fungus steps up to overrule the others in every batch. An anaerobic bacterium (meaning that it is suffocated by oxygen) is found in all of the *hákarl* that has been tested, and a yeast is there too, but the rest of the fungi are all over the place, with a salmagundi of yeasts and molds chewing away in the chunks of flesh. And yet, this is a form of fermentation that is controlled, to some extent, by the draining and drying, which makes me wonder how much worse shark flesh could taste if it decomposed without any management?

Traveling east of Iceland, continuing this foray through the world of rotten fish, we could taste herring, or *surströmming*, in Sweden, which smells like an open sewer; various species of fish dried, salted, and resurrected as *momone* in Ghana; *fesikh* from Egypt; and onward to an extensive menu of fish and fish sauces in Asia.[20] Bacteria and fungi collaborate in the preparation of all of these foods. Some, like *hákarl*, ferment spontaneously; others are inoculated with starters of grains colonized by fungi maintained for this purpose. The resulting foods add flavor to diets, notwithstanding the revulsion experienced by nonnatives, and preserve them from the automatic decomposition that poisons and liquefies fish in the open air. Each of these ethnic foods is a tribute to the resilience and ingenuity of local human populations that strengthened their symbioses with the fungi.

A sauce fermented from fish intestines called *garum* was a staple in the Roman Empire. It was as popular as tomato ketchup and was manufactured outside cities to spare urban residents from its awful smell. Worcestershire sauce includes anchovies and tamarind fruit in its secret recipe and may hint at the glories of *garum* that disappeared with the empire.[21] The history of Worcestershire sauce is as murky as the genesis of the other foods in this chapter, with a tale about a colonist returning from India with a fever for a chutney deflating when a former employee of Lea & Perrins—the British manufacturer—discovered that the gentlemen named by the company had never set foot in the country. With or without any culinary espionage, the recipe for this Indian-inspired sauce was developed in the city of Worcester by two chemists, John Wheeley Lea and William Perrins, and has been sold since the succession of Queen Victoria (1837). Yeasts and bacteria complete the exacting eighteen-month cultivation process, which is a trade secret.

Soy sauce has a clearer pedigree, with scholarly sources converging on its beginnings in China and later introduction to Japan in the thirteenth century by a Zen monk returning from a pilgrimage. The oldest techniques for fermenting soy sauce were probably adapted from recipes for fish sauce to align with Buddhist sutras advocating vegetarianism. Soy sauce is made from a mixture of soybeans and wheat by *kōji*, which is the term for the *Aspergillus* mold that breaks down the proteins in these ingredients and produces glutamate, which gives the sauce its umami or savory taste. Brine is added to the brew for a second-stage fermentation, in which bacteria produce lactic acid and yeasts flavor the sauce as they digest sugars.[22] The Japanese company Kikkoman is the largest producer of soy sauce, which traces its origins to the seventeenth century. Soy sauce has also been produced in Korea for centuries, where other fungi and bacteria engage in a lengthy preparation of seasoned cabbage to create the glorious staple *kimchi*. Fungal foods are powerful emblems of national and regional identity, and their manufacture in other countries has invited claims of cultural appropriation. Producers have engaged in fierce battles over trademarks, and protections have been instituted in many places including the French *appellation d'origine contrôlée* for cheeses and other agricultural products.

As Europeans embraced *Penicillium* as their chief cheese fungus, people in Asia worked with *Mucor* and related bread molds to craft solid foods from beans and cereals, including *sufu* in China, *idli* in India, and *oncom* and tempeh in Indonesia.[23] The consistency of these products ranges from the spreadable *sufu*, which has been compared with cream cheese, to the firm slabs of tempeh. Each of these products is the creation of a complex community of fungi and bacteria that performs a controlled breakdown of proteins and complex carbohydrates to create a more digestible food and adds flavors that range from bland to the demonic. Tempeh has the widest global distribution and is the most overtly fungal of these fermentations in taste and appearance, with whole soybeans glued together by wefts of white mycelium, producing something like an inverted Camembert cheese. It is a popular meat substitute that is especially good when it is cooked after marinating in barbecue sauce and is also served as crispy fritters in East Java called *tempeh kemul*. The reference to ferments with a demonic flavor applies to regional versions of Chinese *sufu* that have been ranked at the far end of a disturbing continuum of tastes that runs from floral to cadaverous and fecal.

The continental differences between the mold species used in the fungal technologies of Europe and Asia are striking and seem to have arisen from the type of food that was fermented rather than the geographical patterns of fungal distribution. All of the fungi were busy with other jobs before domestication: *Penicillium* spores wafting from rotting vegetation were tempted by milk curds in the cool air of France, and other molds growing on animal dung leapt at the opportunity to ferment fresh soybeans in the tropics of Java.[24] The *Aspergillus* recruited to make soy sauce in China was also doing its own thing in nature, feeding on wild grasses for millions of years before its employment by humans. In each case, a combination of local agricultural produce, climate, and human ingenuity produced new kinds of food. There is a fine line between the spoilage and improvement of foods by fungi. Any mold that produced toxins when it started growing on milk curds or sausage skins was sure to be discarded: nobody was going to use it as a starter for the next batch of cheese or cured meat. But when harmless molds showed up with unusual smells they were in with a chance. Some artisans threw them out, but others must have found them sufficiently interesting to give them a try, which is why we have blue cheese in France and salami in Italy. *Hákarl* and *surströmming* must have been born from desperate hunger.

Region by region, over the millennia, the human-fungus symbiosis flourished as people discovered that certain manifestations of moldiness preserved and improved food rather than reducing it to a putrid sludge. Greater control over the outcome of these fermentations was achieved by cultivating fungi on starters that could be mixed into the raw foodstuffs. This stopped other molds from taking hold and spoiling the fermentation. Starters included funky bread for cheesemaking in caves and moldy rice added to cooked soybeans to forge the blocks of tempeh. As the centuries passed, the human-fungus symbiosis deepened, and the technologies were refined long before anyone understood the nature of the strange partners that cobwebbed our food and clouded our drinks.

THE QUORNIAN AGE

Solutions to the mysteries of fermentation began in the nineteenth century, when Louis Pasteur and his contemporaries proved that the bacteria and fungi seen with microscopes were the living agents of food spoilage and brewing. This spurred the deliberate application of microbes in the production of foods and medicines that we call biotechnology. With considerable hubris, a blue road sign declares South San Francisco "The Birthplace of Biotechnology." This claim rests on the accomplishments of the biotech giant, Genentech, which was born there in 1976, but it is worth considering that the Pasteur Institute in Paris was established in 1887, twenty years before South San Francisco was a city. There are, after all, plenty of hotspots in the history of biotechnology. With considerable bias, and no road sign, I argue that Cincinnati is the real birthplace of American biotechnology because this was the site of the first yeast factory after the Civil War.[25] And what about the village of Marlow Bottom in Buckinghamshire? In 1967, a team of food researchers began scouring the planet in search of a mold that would convert starch into protein. Some of the fungi did not grow very well, others grew fine but produced toxins, and then, on April 1, 1968, a species of *Fusarium* was isolated from a compost heap in the village that became the source of the superlative meat substitute trademarked as Quorn.[26] The superlative fungus had been growing under the noses of the investigators, a half-hour walk from their laboratory headquarters.

Today, the Quorn fungus is cultivated in 50-meter-tall fermentation towers, where it is circulated by a column of rising air bubbles and fed with corn syrup, ammonia, and minerals. Energized by the glucose in the syrup, the filaments of this mold reach out and branch into mini-mycelia and concentrate as crumbles of the cholesterol-free foodstuff called mycoprotein. Beginning with one gram of the fungus, each fermentation cycle produces 1,500 tons of Quorn. This is processed into meatless nuggets, sausages, and burgers that are sold in frozen food sections of grocery stores around the world.

Quorn is a very successful product that is enjoyed by those of us who have reduced or eliminated meat-eating and miss the texture of cooked flesh. Tempeh has a similar meat-like structure, but Quorn is a less ambitious food that does not raise any suspicions. Tempeh always makes me think about what I am eating, whereas there is nothing moldy about Quorn at all. It is so good as a breaded nugget that it should, logically, end the torture of broiler chickens. It is ironic that the development of mycoprotein as a tasty substitute for chicken has happened at the same time that chicken breeders and processors have been working toward increasingly tasteless lumps of meat so that the fungus and the animal are practically indistinguishable.

If, however, the urge to eat Quorn derives from the idea that it is environmentally sustainable, we need to think again, because it isn't, or isn't very. And this is the problem with all manufactured fungal protein products and all nonedible fungal products like packaging materials made from mushrooms: they are very energy intensive. Plants are sustainable because they make their own food. Fungi, like animals, must feed on other organisms or their remains. If the Quorn fungus was content eating agricultural waste, that would be a more environmentally benign proposition. It would be decomposing something we cannot digest and turning it into artificial chicken nuggets. But Quorn is raised on corn syrup, which is trucked to the manufacturing plants in heated tanks, and the fungus is propelled around its fermenter stacks using powerful electric motors, fertilized with ammonia, and on, and on. Carbon dioxide billows into the air at every stage of the fermentation process, as it does whenever we make things. We can judge that Quorn is better for the environment than meat or fish, but carbon neutrality is as elusive as the battery-powered car whose only emission is water.[27]

Like the other food molds described in this chapter, the Quorn *Fusarium* is related to fungi that destroy crops, spoil harvested grains, and infect humans. *Fusarium* is a huge genus that encompasses as many as a thousand species, including fungi that live in soil, plant pathogens that attack wheat and barley, and another species that is driving the commercial banana toward extinction. This presents a marketing challenge, which is why we refer to mycoprotein rather than fungus protein, mold nuggets, or mildew meals. Some *Fusarium* species produce toxins (see chapter 8), but the Quorn *Fusarium* has never caused any problems as it swirls in its colossal fermenters.[28] Other molds are also being raised on an industrial scale to produce enzymes that flavor and color fruit juices, tenderize meat, process cereals for animal feed, and prepare ingredients for baking and brewing. Fungal enzymes remove lactose from milk, modify milk formulas for infants, and are used to coagulate milk without using rennet that comes from calf stomachs. Enzymes from filamentous fungi are also used to turn cornstarch into the corn syrup that sweetens sodas, breakfast cereals, candies, cookies, and most of the rest of the inventory of prepackaged junk foods. Incidentally, the syrup used to raise Quorn is produced from cornstarch using *Aspergillus* enzymes, and yeast is grown on the same stuff to manufacture bioethanol.

Fungal biotechnology, which began with palm wine and fermented milk, has evolved into a global industrial partnership between humans and fungi. Fungi have crept into the modern diet, one box, bottle, and aluminum can at a time, until we cannot live a day without them. The economic value of this symbiosis to the economy of the United States exceeds $1 trillion, which is comparable to the annual revenue from the automotive industry.[29]

BACK TO THE WOODS

Fermentation comes from the Latin *fermentare*, which means to leaven or make bread dough rise, and its meaning had been extended to describe bubbling, effervescing, and alcoholic fermentation by the 1600s.

Yeast is breaking down sugars in all these processes. It is doing what comes naturally to every fungus and defines the kingdom—namely, decomposition. So, fermentation describes forms of disassembly, decomposition, or decay that we have learned to control to ensure the formation of useful rather than useless breakdown products. Once we recognize that some form of decomposition drives all of the industrial applications of fungi, the repulsive smell of some fermented foods makes complete sense. Although we do not tend to think about wood rotting as an example of fermentation, the fungi that do this in nature are performing the same kind of chemical transformations as the fungi in blue cheese, soy sauce, and tempeh. For this reason, we might expect mushrooms to be as nutritious as Roquefort cheese or mycoprotein burgers, but we would be wrong. Although the structure of a Quorn nugget is similar to the internal anatomy of a mushroom, Quorn is packed with protein, whereas a mushroom is composed of indigestible fiber and pumped-up with water. Mushrooms are very light on calories because their colonies or mycelia feed on plant debris. Mycelia may come across some scraps of animal protein from time to time and grab some snacks from tree roots if they belong to a mycorrhizal fungus, but the developing mushroom is never going to be flooded with sugars like a mold in an industrial fermenter.

The fruit body arises from the mycelium as a platform for releasing spores. Over the span of 150 million years, the fungi that produce these umbrella-shaped sex organs found ways to make sensibly sturdy structures that support the gills using the bare minimum of energy. This economy was achieved by absorbing water like a sponge to stay upright rather than developing stiffening tissues like the fibers in a plant. With their watery interior, mushrooms are a poor food source for animals. Gram for gram, mycelia have the same energetic value as cheese, whereas gilled mushrooms that grow from them have no more calories than a lettuce. Truffles are a luxurious exception to this rule because they evolved to attract animals for dispersal and are as fattening as Roquefort.[30]

Nobody ever got fat eating mushrooms. Oysters, portobellos, porcini, and other above-ground mushrooms contain a healthy mix of minerals, but nothing that we do not consume in larger quantities in a serving of fruit or a bowl of breakfast cereal. Mushrooms are eaten for pleasure, not for survival. They add texture to recipes and some of them pack spectacular flavors. They are also served in weight-loss diets, including the M-Plan, promoted by several celebrities, which recommends replacing one daily meal with a plate of mushrooms cooked in as little oil as possible. This is a straightforward approach to reducing calories and is no more or less effective than substituting a salad without any dressing for a richer meal. It is also difficult to stick with, and the body tends to oppose the loss of calories in the mushroom dish by stimulating greater food intake in the meal that follows. Part of the appeal of the M-Plan and other diets that emphasize mushrooms comes from the widespread belief that fungi have almost magical powers— always unspecified—that can help reshape the body more than vegetables. This illusion is also part of the allure of medicinal mushrooms, which is the topic of chapter 7.

7

Treating

MEDICINES FROM FUNGI

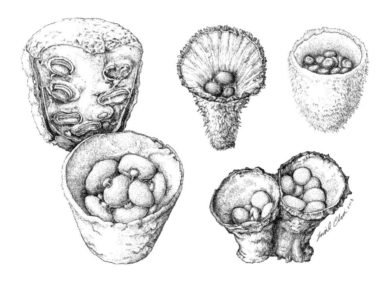

A GERMAN COUPLE hiking in the mountainous border between Italy and Austria in 1991 came across a corpse, half frozen in a glacier and slumped, face down, in a slurry of melting ice. They had discovered the mummified body of a forty-four-year-old man who had died more than five thousand years ago. We call him the Iceman, or Ötzi, which is a reference to the Ötzal Alps where he was found. An arrowhead was lodged in his shoulder blade, and the Iceman's body bore other scars of combat including a head wound. He may have bled to death or died from hypothermia before snow and ice preserved his slender body and belongings. A pair of smooth white cuttings from the flesh of a shelf fungus were threaded on ribbons of leather, and a little pouch contained fragments of another fungus that served as a firelighter.[1] Ötzi was a Neolithic mycologist.

The white trinkets came from a fungus called the birch conk or polypore. Ötzi would have had no trouble finding this mushroom. It is a common parasite of older birch trees, which rots the trunk of the tree before bursting through the papery bark to form a squidgy clod that expands into the shape of a horse hoof. As the fruit body matures, a layer of tiny vertical tubes develops on the bottom of the hoof that opens into thousands of tiny holes—*poly-pores*. Pursuing a circadian rhythm, with maximum activity during the nighttime, the fungus drops millions of spores per hour from these pores. It grows best on trees weakened by drought or stressed by overcrowding, and the birches are left so frail that they topple with a little shove.

In 1998, an Italian anthropologist studying the mummy proposed that Ötzi had used "measured doses" of the fungus to induce "strong though short-lived bouts of diarrhea."[2] This was a remedy, he said, for internal parasites, whose eggs were identified in Ötzi's mummified rectum. Treatments for intestinal worms would certainly have been valuable in Ötzi's time, when the human gut was a more festive arena than today's plumbing system, with (to the tune of "My Favorite Things") *Roundworms in most guts and hookworms in plenty / Segmented tapeworms that make you feel empty / Many amoebas and pinworms like strings / These were a few of our nightmarish things.* Ötzi had one of the commonest parasites, the human whipworm, which is a roundworm or nematode that causes painful bowel movements and loss of appetite in heavy infections.[3] Few readers will have encountered this parasite, but everyone in Ötzi's tribe would have known them.

But the case for Ötzi's use of the birch polypore as a worm treatment is very flimsy. Diarrhea is a symptom of worm infestation rather than a treatment, although temporary relief might come from emptying the bowel. There is no evidence, in any case, that the birch polypore works as a laxative, nor that the Iceman measured his doses of the fungus. However, these baseless ideas were published in a top-tier medical journal, *The Lancet*, and attracted a lot of media attention. With the imprimatur of this periodical, few readers questioned the credibility of the medicinal claim, and Ötzi became idolized as an aboriginal apothecary who trudged through the snow to bear witness, in mummified form, to the ancient wisdom that we have lost.

Misrepresentation afflicts the whole field of research on medicinal mushrooms, making it difficult to tease out fact from fiction, but this is the brief of this chapter. Although the birch fungus does not seem to have been used by anyone as a worm remedy, it was adopted as a medicine to treat other conditions. It appears in the literature on traditional medical practices in Russia and Central Europe as an antiseptic and styptic (to stop bleeding). The birch fungus is also mentioned as an anti-inflammatory and anticancer agent and, more specifically, as a veterinary cure for vaginal tumors in dogs.[4] Mushroom poultices have certainly been useful for treating minor wounds, but I bet that Ötzi kept the conk for a different purpose. A polypore species that grows on willows was carried by Blackfoot people, or Niitsitapi, and members of other nations of the Northern Plains in the same way as the Iceman, who carved white beads from the fruit bodies and hung them on rawhide thongs.[5] Museum collections in Alberta, Canada, include Blackfoot charms made from human scalps decorated with these white beads. They were also sewn onto sacred robes and worn on necklaces. One archival photograph from the early twentieth century shows a chain of the fruit bodies around the neck of a horse, looking like the modern rhythm beads familiar to equestrians. Native Americans seem to have imbued their fungus with spiritual significance, and Ötzi may have revered the birch polypore for similar reasons. This is a more logical interpretation than creating a story about the deworming properties of a fungus from thin air.

MUSHROOM THERAPIES

Illustrations of mushrooms in Neolithic petroglyphs are expressions of the earliest conscious relationships between humans and fungi. We do not know the significance of fungi to these people, but the psychoactive properties of some species suggest that they would have featured in animist religions (this is discussed further in chapter 9). Wherever mushrooms were plentiful, people would also have learned to avoid the poisonous kinds, cook the tastiest ones, and gather a few special fruit bodies as medicines. Although the practical uses of the fungi diversified, they never escaped their association with the supernatural. This blurring of the distinction between medicine and magic continues today and is the foundation of the multibillion-dollar market for mushroom extracts.[6]

The medicinal mushroom business depends on fewer than a dozen fungi that are advertised as therapeutic stars. Reishi or lingzhi is a bracket fungus with a polished red surface; shiitake is an unassuming brownish umbrella with gills; maitake, or hen-of-the-woods, grows at the base of old trees as an outpouring of crowded gray flaps; and turkey tail erupts from decomposing logs as thin fans patterned with vivid stripes. Three more will cover the "medicinal seven": cordyceps fruits from dead caterpillars as a firm pencil-sized stalk; chaga erupts from wounded birch trees (same as Ötzi's conk) before splitting and drying into a charcoaled lump; and lastly, lion's mane matures as a rounded wodge of pure white spines that looks like a frozen waterfall.[7]

All of these fungi are described as medicinal mushrooms, and none of them have proven medicinal effects. Devotees of medicinal mushrooms will disagree with this statement, so it is worth restating: many people believe that mushrooms are useful for treating illnesses, and some of these claims may be true, but they are not backed up by dependable scientific evidence. Mushrooms and mushroom extracts are not like other drugs, because they are treated as foods rather than medicines in the United States and are sold as dietary supplements and herbal remedies. This means that they escape the stringent testing and regulation applied to over-the-counter medicines, including painkillers and cough treatments and the drugs prescribed by doctors. The absence of regulation may seem refreshing for people tired of government interference in their lives, but this leaves us at the mercy of the businesses that market medicinal mushrooms. For all their faults, the major pharmaceutical companies are forced to follow some rules and are self-disciplined by the continuous threat of lawsuits by consumers.

Medicinal mushrooms are sold as slices of dried fruit bodies and powders and, with no requirement for the manufacturers to identify the active ingredients, we cannot assess the potency of one product line relative to another. Imagine buying a bottle of aspirin and learning that the pills contained nothing but chalk. We would be reasonably upset, yet a shameless company could fill capsules with cornstarch mixed with a pinch of anything mushroomy and sell them as cordyceps supplement with no legal repercussions. This is not an exaggeration: there are no industry-wide standards for assessing the ingredients in medicinal mushroom products.

Web postings proclaim medicinal mushrooms as wellsprings of "health-boosting vitamins, minerals, and antioxidants" and "nutrients that support the body's natural immune functions and balance."[8] The term "well-being" tends to crop up a lot and is as difficult to define as "nutraceutical," with which it is often associated. The language of the industry is demeaning to anyone with a modicum of intelligence, but most of us are predisposed to wishful thinking when it comes to health issues. It is easy to feel aloof when we are feeling tiptop, not so much when we are, indeed, drifting into the arena of the unwell. As long as the consumer does not favor mushrooms to the exclusion of life-saving prescription drugs, these potions are harmless, and the remarkable power of the placebo effect can be worth the purchase price.

LION'S MANE AND BETA-GLUCANS

The lion's mane mushroom, or hedgehog fungus, offers a useful case study for a particular species. The mycelium of this fungus feeds on dead trees and can keep fruiting for decades from the same log until the wood dissolves into the forest floor. The mushroom is a naked spore factory that operates as an inside-out polypore by discharging spores from the surface of its spines. Its medicinal status rests on the experimental effect of extracts from the fruit bodies on the growth of isolated nerve cells in culture dishes. Rats fed the powdered mushroom have also shown increased levels of nerve growth factor (NGF), which is a chemical messenger that contributes to the maintenance of the nervous system.[9] The chemicals in lion's mane that are responsible for these effects are called hericenones and erinacines, referring to the Latin name of the fungus,

Hericium erinaceus, meaning hedgehog-hedgehog. In addition to the research on cultured nerve cells and rats, a human study from Japan examined the effect of daily doses of powdered lion's mane on patients diagnosed with mild cognitive impairment.[10] After sixteen weeks of treatment, patients given mushroom capsules improved their scores on a standard screening test for dementia relative to controls.

These investigations are an interesting starting point for more research, not for marketing a mushroom as a medicine with proven curative powers. Tero Isokauppila, author of *Healing Mushrooms*, is less cautious. He says that lion's mane "can potentially reverse the cognitive deterioration that creeps up on all of us as we age," which is supported by the Japanese study, and goes on to describe the successful use of the mushroom in healing brain damage.[11] Paul Stamets, founder of FungiPerfecti, "Makers of Host Defense® Mushrooms," is similarly excited about this species. His website says that a daily dose of the fungus, in the form of capsules, "optimizes nervous and immune system health."[12] In business since 1980, FungiPerfecti adds the following caveat to its endorsements of nutritional supplements: "This product is not intended to diagnose, treat, cure or prevent any disease."

Less restrained promoters of medicinal mushrooms declare that lion's mane "prevents neurodegenerative diseases like Alzheimer's and Parkinson's"; is a treatment for depression and anxiety, digestive ulcers, and cancer; reduces the risk of heart disease; and helps manage diabetes. According to some websites, lion's mane cures erectile dysfunction, while contrary sources of online wisdom declare that the mushroom reduces libido.[13] The truths and fabrications about lion's mane illustrate the way that a scattering of scientific experiments supporting the pharmacological effects of a fungus become stretched into fantastical declarations about mushrooms as the universal panacea— the discovery of the philosophers' stone that eluded the alchemists of the Middle Ages.

The nerve agents isolated from lion's mane have not been found in other mushrooms, whereas another group of medicinal compounds are common as muck, loaded into the cell walls of every fungus. These are the beta-glucans, which are components of the hyphae that form mycelia and the squishy flesh of mushrooms. Beta-glucans also come from oats and barley. Glucans are polymers of sugars or polysaccharides. Cellulose, which is made by plants, is another polysaccharide. Molecules of cellulose are assembled as thin strands, which are packed lengthwise to form fatter threads, like the wires inside electrical cables. Cellulose passes through the gut intact, where it serves as the bulk of the indigestible fiber in our diet. The sugars that form beta-glucan molecules are connected in a different fashion, which allows the smaller ones to dissolve in water, making them part of the soluble fiber in the diet. These smaller molecules bind to the surface of macrophages and other cells involved in the innate immune defenses that are stationed in the wall of the intestine.[14]

Beta-glucans are recognized by the immune systems of all animals from butterflies to bison on dry land and crabs to cetaceans at sea. Even the sponge ancestors of the whole animal kingdom are irritated by beta-glucans.[15] This universal response to beta-glucans speaks to the supportive relationships that have been forged between fungi and animals from the evolutionary get-go, as well as the mortal threat posed by fungal infections. The human immune system recognizes beta-glucans as a measure of the status of the fungi on the body along this spectrum of interactions. It is attentive to beta-glucans released from yeasts and other fungi passing harmlessly along the gut without provoking an inflammatory reaction. A more aggressive response is mounted when larger pulses of beta-glucans are detected from a fungus that is damaging our tissues.

From the foundation of beta-glucan science, followers of alternative medicine have turned these sugary strings into the mycological equivalent of ass gelatin—a traditional Chinese balm, also known as donkey-hide glue, which is a miracle treatment for cancer. Ass gelatin does not, of course, cure any forms of cancer, but that has not interfered with its traditional uses. The positive thing about beta-glucans is that they do have proven effects on the immune system, which is one of the few reliable facts that can be established about medicinal mushrooms. This immune activity has led investigators to pay serious attention to beta-glucans, and there have been several controlled trials to see if they might prove useful as cancer therapies.

The traditional applications of mushrooms for treating cancer are provocative, but the immunological logic is very weak. Why should a chemical compound that stimulates the immune system to combat fungi turn our defenses against our own cells that have become cancerous? The most hopeful answer is that an immune system aroused by any stimulus might be better at recognizing and clearing cancer cells than a weakened immune system. Slender support for this idea comes from the effect of beta-glucans as adjuvants—compounds that increase the action of another medicine. Lentinan, which is a preparation of beta-glucans from shiitake mushrooms, has been used in the treatment of stomach cancer in Japan for many years.[16] Coupling conventional chemotherapy with lentinan injections appears to increase the survival time of patients for an average of four months, and similarly supportive effects have been reported for the treatment of lung cancer in China.[17] There is some hope here.

Like the work on lion's mane, the preliminary studies on lentinan from shiitake encourage further research. But they do not justify the astounding successes of medicinal mushrooms imagined by their disciples, according to whom, beta-glucans are "considered the best immune modulators globally," and are well on their way to being acclaimed as "a 21st Century Miracle."[18] If beta-glucans turn out to be beneficial, this would reinforce the regular consumption of mushrooms as part of any healthy diet. Mushrooms certainly add flavor and texture to our meals without many calories (see chapter 6), but grocers have refrained, so far, from advertising jars of sliced buttons or fresh portobellos as the secrets to longevity. Dr. Djibril Ba, from the College of Medicine at Penn State, is more bullish, suggesting that eating the fruit bodies of any fungus lowers the odds of dying from any cause.[19] His evidence comes from a cohort study in which participants answered questionnaires between 1988 and 1994, and their fate—dead or alive—was assessed twenty years later. The survivors were more likely than their unfortunate peers to have reported eating some kind of mushroom in any quantity and in any form in the twenty-four hours before they were questioned. In principle, this means that someone who ate a bowl of mushroom soup on the day before they filled out a questionnaire, or said that they did so, tended to live longer than someone who did not, or did not remember doing so. The proposed effect was modest, meaning that mushroom eaters were 14 percent less likely to die from all causes, which would include mortal illnesses and, one imagines, falling off ladders, gunshot wounds, suicide, and shark attacks. The figures from this study suggest that one in five fifty-year-old men who ate mushrooms at the beginning of the study died before age seventy, compared with one in four who did not eat mushrooms. Dr. Ba has also scoured through epidemiological data and found that mushroom consumption is associated with a lower risk of developing cancer and suffering from depression.[20]

If eating fruit bodies is as beneficial as Dr. Ba claims, he will receive a Nobel Prize, and mushrooms and every food containing fungi will be marketed as the elixirs of life, which, come to think of it, is exactly what purveyors of medicinals have been doing for years. However, what the cohort studies actually show is that there seems to be a weak relationship between eating mushrooms and staying alive, but this does not mean that eating them has any direct effect on longevity or happiness. Something about the lifestyles of the mushroom eaters influenced their survival, or something about the people who did not eat mushrooms influenced their demise. Regarding the proposed antidepressive effect of eating mushrooms, a Polish investigator, Piotr Rzymski, suggested that the enjoyment of foraging for wild mushrooms might explain the promotion of mental health in Ba's study.[21] This seems unlikely, given the rarity of mushrooming among Americans, but it points to the confounding variables in a study of this kind and the need for caution in interpreting cohort studies. If mycophobes rode motorcycles more than mycophiles, this would explain the protective effect of the fungi. Pick other less risky variables in human behavior and we might find faint statistical signals of well-being among people who do crossword puzzles or keep goldfish.

Students of medicine and journalism would benefit from reviewing and discussing cohort studies like this as lessons in critical thinking and scientific objectivity. Something interesting could come from Dr. Ba's investigations on mushroom eating, but to accept the notion that mushrooms cure cancer would be an act of naivete comparable to belief in the

Easter Bunny. For an example of good practices in mushroom medicine, students should consult an authoritative review published in the *American Journal of Medicine* in 2021.[22] The authors of this study analyzed almost 1,500 published studies on the effects of mushrooms on cardiovascular health and concluded that only seven of them, or 0.5 percent, were sufficiently reliable to recommend further investigation.

OBJECTIVITY AND OPTIMISM

A few years ago, I published an article titled "Are Mushrooms Medicinal?" in the journal *Fungal Biology*, which was downloaded more than any previous paper in the history of the periodical.[23] Supportive emails arrived from fellow scientists, but a much larger number of messages questioned my intelligence and motives. One critic asked, "Did you even do a lit review on the topic before you wrote this paper?" and continued, "It just seems like you're being willfully ignorant, intentionally misleading, or have an interesting criteria [*sic*] for what constitutes validity." My criteria [*sic*] is, show me the evidence, which is codified in Hitchens's razor, or epistemological rule, authored by the late journalist Christopher Hitchens: "What can be asserted without evidence can also be dismissed without evidence."[24] Another correspondent was perplexed by my judgment and suggested that I read their patent applications.

Here is the first sentence from my contrarian essay: "Despite the longstanding use of dried mushrooms and mushroom extracts in traditional Chinese medicine, there is no scientific evidence to support the effectiveness of these preparations in the treatment of human disease." And here is the last: "It is time to treat anti-aging tonics made from mushrooms as a sad phase in the history of mycology and proceed with the exploration of novel compounds with the potential to change the course of our modern plagues."

The importance of mushrooms in Chinese medicine is a good argument in favor of their usefulness in some applications. If there is no evidence for their value in treating illnesses, why would they have been endorsed by practitioners for so long? Part of the difficulty in answering this question lies in the philosophical gulf between Chinese and

Western medicine. We know that shiitake mushrooms have been cultivated on logs in China for centuries and have been prescribed for re-
plenishing *qi*, which is translated as the vital energy that flows through
the body. This concept overlaps with the humors of the Greek physicians, which held sway in European medicine for two thousand years
and were mentioned in chapter 2. Now we demand evidence of changes
in blood sugar if a drug is used to treat diabetes, or a drop in blood pressure after swallowing an antihypertensive medicine.

The continuing claims made for the uses of mushrooms should
stretch anyone's credulity. To illustrate, I will search for "shiitake" combined separately with each of the following illnesses and health conditions starting with the letter a: acne, AIDS, Alzheimer's disease, anthrax,
arthritis, asthma, and autism. Shiitake is endorsed as a treatment for
all of them.[25] Pick a different letter and see for yourself. Inspired by
Mr. Hitchens, I offer my own razor:

> A cure alleged for everything is an effective treatment for nothing.

By deifying shiitake and other species of fungi, the medicinal mushroom magicians—the champions of champignons—betray themselves
as con artists to anyone willing to ask a few questions. Some of these
latter-day shamans are deliberate scammers, others know no better, and
I am aware that I am howling at the moon. Belief in the baseless powers
of medicinal mushrooms may also be strengthened by the widespread
skepticism toward medical authority.

All of this is unfortunate for the study of mycology because mushrooms are probably brimming with undiscovered medicines that might
be developed using modern pharmacological methods. Why then, one
might ask, has there been such a rich history of drug discovery in plants,
while most of the woodland fungi have been ignored? Aspirin, ephed-
rine, opiates, and quinine were used by herbalists in the form of unre-
fined preparations from plants for thousands of years before any attempts
at purification. The historical success of these plant extracts in medicine
argues for a similar impact of mushrooms, but confusion about their
nature ("these bastard plants, or excrescences") and toxicity ("some are
very venomous and full of poison") meant that they were neglected by

the great herbalists of Europe. The quotes come from the English herbalist John Gerard (1545–1612), who endorsed the attitude of the Roman
poet Horace toward fungi: "*Pratensibus optima fungis natura est* [mush-
rooms from the meadows are best]; *aliis male creditur* [others are not to
be trusted]."[26] This advice comes from his *Satires*, which served as a
Roman self-help guide to a life of contentment and reflected the danger
of misidentifying mushrooms. Chinese healers were more adventurous
and became experts in identifying the wild mushrooms that they used
to treat illnesses. But without the Western transition from herbalism to
pharmacology, practitioners of traditional Chinese medicine made no
attempt to purify individual compounds from fungi. They saw no need
to do so, and the disjunction between traditional and Western medicine
endures today.

These are some of the factors that explain the absence of pharmacological interest in the fungi until the twentieth century. This means that
we are playing catch-up and is why we need to apply the most up-to-
date methods to explore this natural pharmacopeia.[27] The theoretical
argument for drug prospecting from fungal fruit bodies—namely, that
they are likely to be there—is bolstered by the plethora of powerful
medicines that have already come from fungi. The first antibiotic, penicillin, from an English strain of *Penicillium*, was discovered in the 1920s;
a second antibiotic called cephalosporin was isolated from a Sardinian
mold in the 1940s; and cyclosporin, which prevents organ rejection after
transplants, came from a Norwegian fungus in the 1970s. Lovastatin, a
cholesterol-lowering drug isolated from an *Aspergillus* species, entered
clinical trials in the 1980s, and a variety of antifungal agents isolated
from one fungus to treat infections caused by another fungus are also
synthesized by molds. The oldest fungal medicine is ergotamine, which
is produced by the ergot fungus that infects cereals and has been prescribed as a treatment for migraines for more than a century. (Ergot
poisoning is featured in chapter 8.)

All of these medicines come from the feeding colonies of molds
rather than mushrooms, although some of them are manufactured in
the lab today without any participation by a living fungus. Lovastatin
from *Aspergillus* has also been found in oyster mushrooms, but at levels

that are too low to have any effect on cholesterol levels in the bloodstream. The search for mushroom medicines is complicated by differences in chemical activity between the fruit bodies and their supporting mycelia.[28] These develop because the fungi balance their metabolic processes between the feeding mycelium and the reproductive mushroom to sustain both phases of the life cycle without wasting energy. The way forward in the search for medicines from mushroom-forming fungi—from their fruit bodies or supporting mycelia—lies in the study of genomics. By sequencing and manipulating whole genomes of fungi, we can search for the genes that control the formation of useful chemical compounds. If we want to find new sources of antibiotics that belong to classes of chemicals that are already known to destroy bacteria, we can unearth them by analyzing little pieces of a mushroom. Even if the mushroom does not make any of the antibiotics, the fruit body will contain the genes for doing so if its mycelium is chugging them out in the soil.

Finding the instructions is difficult because they are buried within the tens of billions of As, Ts, Gs, and Cs in the fungal DNA. The workaround involves breaking up the genome into hundreds of scraps called artificial chromosomes and seeing what each segment of DNA can do on its own.[29] This technology is known as genome mining. It allows investigators to pore over the DNA of a fungus, strand by strand, like pirates pulling necklaces from a treasure chest. Rather than selecting mushrooms randomly, researchers can be guided by information on the traditional uses of mushrooms.[30] If Eastern Europeans have been using a particular bracket fungus to treat rheumatism for centuries, this is a good candidate for drug prospecting. Even if chemists have failed to find active chemicals in earlier studies, the fungus may be worth a second look because its DNA may harbor the instructions for making something useful. Genome mining has the power to transform the medicinal uses of mushrooms into a science.

While mushroom medicine has gone nowhere, the yeast used for brewing and baking has transformed the pharmaceutical industry as an infinitely pliable platform for drug production. Ordered to make human insulin using the human gene, yeast complies and produces half of the global supply of injectable insulin to combat diabetes. Modified with a DNA sequence from the human papilloma virus (HPV), yeast translates this into copies of the protein that forms the shell of the viral particle. Separated from the yeast, this protein is injected as a vaccine against HPV, which has the potential to eliminate the form of cervical cancer linked to this infection.[31] Another medicine from genetically modified yeast is used to treat an age-related eye condition, and there is a lot of research on using yeast cells to synthesize painkillers. When we consider antibiotics from molds and all these drugs manufactured by yeast, it is evident that fungi are an indispensable source of modern medicines. To add new compounds from fungi to the pharmacy, we need to bring the chemistry of mushrooms into the mainstream.

FROM THE NEOLITHIC TO THE ANTHROPOCENE

Ötzi's world was filled with mythology and mycology. The Iceman's woods and meadows were a mycological wonderland, and his childhood would have been filled with stories about toadstools, and witches, and wood spirits. We will never know why Ötzi carried the birch conks, whether they were charms, or if he valued them as medicinals. As a plentiful and diverse resource, people would have put fungi to use in one way or another, even if they learned to steer clear of most of them for fear of poisoning. All we can say with any confidence is that some mushrooms were handy first aids for treating wounds and that others were sought for their hallucinogenic properties (which are described in chapter 9). The challenge in the twenty-first century is to reengage with the therapeutic possibilities of mushrooms without resorting to superstition in treating disease. Science needs to work alongside a sense of mycological curiosity to find the pharmaceutical riches in the woods.

My confidence in the abundance of undiscovered mycological medicines is increased by the chemical makeup of some of the least conspicuous mushrooms. Smaller than thumbnails, beautiful as any flower, bird's nest fungi sit on broken twigs waiting for raindrops. Drops falling into the cups flick the spore-chocked eggs into the air. Each discharged egg drags a sticky harness—think tiny chameleon tongue—that wraps around a grass stalk, bringing the fungus to a halt after a one-second flight.[32] The discovery of antibiotics in these acrobatic species began with a chance observation made by Alex Olchowecki, a graduate student at the University of Alberta in the 1960s.[33] Alex noticed that bacteria contaminating a culture of one of these fungi were wiped out from the region closest to the growing hyphae. A similar bacterium-free halo was observed around hyphae of *Penicillium notatum* by another Alex, Alexander Fleming, in 1928. Fleming's discovery of penicillin led to a Nobel Prize in 1945, shared with Howard Florey and Ernst Chain, who did the hard work in antibiotic development. Olchowecki did not receive any prizes for discovering the antibiotic from the bird's nest fungus, called cyathin, but went on to enjoy a distinguished career at the University of Manitoba.

Work on the chemistry of the bird's nest fungi has continued, and a whole class of antibiotics belonging to the "cyathane diterpenes" has been found in these pretty little mushrooms. None are as powerful as penicillin at killing bacteria, but some of these molecules stimulate the growth of nerve cells, like the drugs isolated from lion's mane.[34] Cyathanes purified from bird's nests also work against cultures of cancer cells. The detection of antibiotics in the bird's nest fungi increases the odds that some useful molecules are hiding in fairy bonnets and eyelash cups, pinwheels, violet corals, orange jellies, bleeding mycenas, and the rest of the woodland fungi. Now is the time for investigators and investors to get serious: *carpe boletum*, as Horace would have said, if he had developed a more favorable opinion of mushrooms. On the other hand, the search for medicines in fruit bodies is likely to reveal as many toxins as treatments, and we consider the ruinous nature of some of the prettiest mushrooms in chapter 8.

8

Poisoning

TOXINS IN MUSHROOMS
AND MOLDS

LIKE OTHER FACETS of the relationship between humans and fungi, the presence of fungi in the diet and the uses of medicinal mushrooms are mixtures of conscious and unconscious interactions. We have seen how fungi are front and center in the fermentation of cheeses and staple foods in Asian cuisine, and that molds have become the "meat" of mycoprotein nuggets. Beer brewing, winemaking, and bread leavening by yeast are other obvious examples of the hidden role of fungi in our diet. But eating mushrooms, whether for food or medicine, is the most conspicuous part of human mycophagy and has assumed such significance in human culture that it is the first thing that most people consider in our associations with fungi. In this chapter we look at mushroom poisoning and our equally unintentional exposure to the toxins produced by molds.

On Easter weekend in 2020, Dr. Anna Whitehead, a physician in New Zealand, picked some mushrooms beneath an oak tree and cooked them with fresh fish for lunch. She said that she had planned to check what they were but became distracted by work and sautéed slices of the fruit bodies without thinking. She awoke early the next morning and began vomiting green liquid. Suspecting what might have happened, she staggered upstairs and searched for images of poisonous mushrooms on her computer. "Immediately, a picture of death cap mushroom[s] flashed up. I recognized it instantly as the type of mushroom I had picked and eaten."[1] She rang for an ambulance. After a day in hospital hooked to an IV in her arm to keep her hydrated, the symptoms subsided, and she went home. Disaster averted? Not quite.

A few hours later the nausea returned, worse now. The classic honeymoon period associated with death cap poisoning was over, and she returned to the hospital. Toxins from the fungus had been circulating in her bloodstream, killing cells in her liver. The pain in her abdomen was agonizing. She seemed to be dying. But her doctors and nurses pulled a Hail Mary by resuming IV support, and after two days in a critical care unit her liver began to recover. Dr. Whitehead had dodged the reaper. In interviews she said, "I have never ever felt so terrible," far worse than the side effects of the chemotherapy she had received to treat cancer long before the poisoning. She also remembered the strong flavor of the pale green caps, which she had thought was the way wild mushrooms are supposed to taste. Other survivors of poisoning have said that the death cap is the most delicious mushroom they have ever eaten. If a physician like Dr. Whitehead can make an almost fatal error, what hope is there for the rest of us?

A GUIDE TO EDIBILITY

Poisonous mushrooms, sometimes distinguished from the harmless species with the name toadstool, seem a very remote threat in the twenty-first century. They are more likely to be associated with fairy tales about witches in the woods than part of a liberal education. But with a renaissance in interest in foraging for wild food, anyone who plans to eat wild mushrooms needs to pay attention.[2] The death cap, *Amanita phalloides*, deserves special notice because it is an invasive species that has spread from Europe across the world and is poisoning people who mistake it for native species that are edible.[3] We must think carefully before eating any fruit body that we find in the wild. My recommendation for safe mushrooming is to limit foraging to a selection of the tastiest species—to become intimate with their appearance and leave the rest of them to get on with the job of being mushrooms.

When the weather has been warm and wet at the end of the summer, the "savory seven" in the midwestern United States, overlapping with the "medicinal seven" in chapter 7, begins with oyster mushrooms, jutting from logs like shellfish from rocks, but with the taste of a delicately perfumed version of a white button mushroom rather than the briny minerality of a fresh bivalve.[4] This subtle oyster-mushroomy flavor is easily lost by overcooking, but nobody would eat them raw. Lion's mane, the medicinal mushroom, tastes much the same. Young puffballs with pure white innards are gill-less buttons with nothing to offer the epicure besides the surprise of serving them as cooked discs on a pizza, or in any other dish where cultivated mushrooms are expected to appear. Chicken of the woods and maitake are a nudge more interesting in the kitchen: firmer than oysters, they carry a fruity or woodsy fragrance that works well in soups and stews. And, more flavorful than their peers, fruity chanterelles and earthy-nutty porcini rise above the butter and garlic absorbed by their flesh if they are not overcooked. There is an obvious seasonality to this list, with morel species replacing these edibles in the spring, but it is difficult, though not impossible, to confuse any of these mushrooms with poisonous species.

This pedestrian advice will dismay more adventurous mushroomers who lionize false morels, *Gyromitra* species, which contain a toxin that is converted into rocket propellant unless it is boiled away before eating; edible webcaps that are difficult to distinguish from deadly poisonous ones; and even a few kinds of benign amanitas—grisettes and blushers—whose doppelgängers include death caps and the aptly named fool's mushrooms and destroying angels.[5] Serious mushroomers are so deeply invested in identifying fungi that they are unlikely to make mistakes, but the rest of us should be very careful. By flagging the minority of truly deadly mushrooms with skull and crossbones symbols, guidebooks and websites can leave the naive mushroomer with the impression that most of the other mushrooms are edible. Although this is tru-ish, edibility does not equate with palatability. The delicious ones are as scarce as the poisonous, and the taste of most mushrooms varies from chewy-cardboardy to soggy-cardboardy. Safety lies in highlighting the appetizing and unmistakably harmless mushrooms, rather than encouraging people to abandon themselves to the carnival of fruit bodies of varying shapes, sizes, colors, smells, and edibility. Fish are like mushrooms: a few are delicious, some are poisonous, and most make dismal meals and should be allowed to get on with being fish.

The savory seven is adjustable for other regions, although errors occur when a favored mushroom happens to resemble a lethal species like the death cap. None of my midwestern treats look like anything toxic, but the situation is different in Asia, where an assortment of mushrooms collected from the surrounding forests are sold in local markets. These delicacies in southeast Asia include species of *Amanita* with pale yellow caps and cream caps. Other than these subtle differences in color, these fungi have all the hallmarks of the mushrooms called destroying angels—white gills, ring flopping down the top of the stem, and the bottom of the stem stuck in a cup. This explains why unwary foragers familiar with the edible amanitas in Asia are tricked by poisonous ones in North America.[6] California mushroomers have made similar errors in confusing death caps for Pacific amanitas, known as coccora or coccoli, which have a fishy smell and can be substituted for seafood in ceviche.[7]

ALPHA-AMANITIN

The deliberate substitution of lethal for edible amanitas is chronicled in the story of the assassination of Emperor Claudius by his wife, Agrippina, in AD 54. The most satisfying version of the plot begins with Caesar's mushroom, which is an edible amanita. This orange-capped mushroom is eaten in its egg form, known as *ovolo buono* in Italy, before the fruit body emerges and expands like an umbrella. Claudius adored this mushroom, which made death caps the perfect murder weapon. There are other readings of his death, but this one provides the most satisfying end to this disgraceful tyrant.[8] Caesar's mushroom is a rare example of a mushroom that tastes best raw, served in thick slices dressed with a little olive oil and lemon juice, and is so sought after that it has been given protected status to prevent over-collecting.[9]

While most poisonings result from failures in mushroom identification, the medical literature includes a few cases in which people have knowingly eaten death caps in suicide attempts. A sad case in Italy involved a young woman who had learned a lot about mushrooms from her father, who was a keen amateur mycologist, collected three large death caps and made sure that she ate a lethal quantity.[10] She would have died but was rushed by her parents to the hospital, where she was saved with a liver transplant. A stranger story of mushroom poisoning involves a bizarre case of experimentation by a Turkish man who ate two death caps to determine whether they were safe: "He told the household that if nothing happened to him, they could eat the remaining mushrooms together the next day."[11] Hours after his meal he developed severe symptoms of gastrointestinal distress and, after some resistance, was taken to the emergency room by his family. Once in the hospital he responded to rehydration and recovered after a few days. This was a remarkable comeback because he had consumed three times the fatal dose of death caps.

After the flesh of a death cap dissolves in the stomach, its toxins are absorbed from the gut and circulate around the body.[12] The worst of the poisons is alpha-amanitin. Amanitin interferes with an indispensable enzyme that ratchets itself along DNA strands, reading and transcribing the genetic code in the first step of protein synthesis. Cells exposed to amanitin shut down without a continuous supply of proteins, and the liver is wrecked as it concentrates the toxin from the bloodstream. Amanitin can be mopped up by drinking a jet-black slurry of activated charcoal immediately after eating the mushrooms. This remedy is useless after the toxin reaches the small intestine and the poisoning symptoms proceed. Once this happens, the best treatment is intravenous therapy to maintain hydration and give the body a fighting chance to flush the toxin away in the urine. Poison that is not filtered by the kidneys is returned to the bloodstream, where it recirculates and pummels the liver afresh. Amanitin is one thousand times deadlier than aspirin.[13] Experimental treatments include dialysis to assist the natural action of the kidneys. Some physicians also suggest draining the bile duct, which conveys the toxin from the liver and gall bladder into the small intestine, to help eliminate the poison. Evidence for the effectiveness of these treatments is limited. The same uncertainty applies to the use of high doses of penicillin to increase the excretion of amanitin in the urine, and to silymarin, a drug extracted from milk thistle plants that offers some protection to liver cells.

Decades of alcoholism do not come close to matching the acute liver damage resulting from eating a single death cap mushroom, and an organ transplant is the only option when the liver does not rebound. Transplantation is followed by lifetime support to prevent rejection and this, ironically, can be provided by two drugs that come from molds: cyclosporin produced by a soil fungus (see chapter 7), and myophenolic acid (MPA) from a *Penicillium*. The *Penicillium* that produces MPA grows on mudflats, sand, stored fruit, wood, and the surface of decomposing mushrooms. This offers a circular mycological meditation, from a disastrous woodland foray to a liver transplant and on to a life-sustaining treatment with a medicine produced by a mold that grows on mushrooms: from fungal illness to fungal cure.

While the death cap is an expert in liver damage, Smith's amanita attacks the kidneys. This renal specialist grows in the Pacific Northwest, which presents a problem because it is mistaken for the delicious matsutakes, which have a cult following in the region.[14] Most victims recover kidney function a few weeks after their encounter with this species, but this poisoning syndrome is another reason to be wary about eating any mushroom wearing a ring and sitting in a cup. This does not mean, un-fortunately, that the cupless and ringless mushrooms are safe. Far from it. Amanitin is also produced by the autumn skullcap, which is a little brown mushroom.[15] This is a species of *Galerina* that decomposes wood and produces clusters of fruit bodies and has been eaten by unwary shroomers who think they have found psychotropic psilocybes. Skull-caps are cupless mushrooms that can come with or without a thin ring on the upper part of the stem, making a mockery of any simple rules for recognizing toxic fungi.

ALL MUSHROOMS ARE POISONISH

Although alpha-amanitin causes most of the serious poisonings, the list of problematic mushrooms goes way beyond fruit bodies contain-ing this compound.[16] Thirty or more species of webcaps contain orel-lanine, which is a toxin that targets the kidneys. The symptoms of web-cap poisoning can be obscured by a honeymoon period, like the delay after eating death caps, but continuing for weeks before the injuries become evident. In other mushroom poisonings, the toxins have not been identified. Yellow knight poisoning is the best example of this kind of cryptic toxicity, in which something in the fruit bodies causes muscle damage. The yellow knight is a flat-capped mushroom that grows in coniferous forests and has been eaten in many European countries for centuries. It has been fried, boiled, and pickled, listed as a favorite in mushroom guides and cookbooks, and was celebrated as a wild food without any concern for its safety until clusters of poisonings were reported from France, Poland, Lithuania, and Germany. The first dozen cases were admitted to French hospitals between 1992 and 2000.[17] Creatine kinase levels in the bloodstream of the patients soared, which was a marker for muscle breakdown, and their urine darkened with proteins and blood cells. Patients felt nauseous, sweated profusely, and experienced leg weakness that left some of them unable to walk. Most recovered within a couple of weeks, but three of the patients died.

What happened here? Yellow knight has a lot of look-alikes, so it is possible that the poisoning was caused by a different species of mushroom. But there is an alternative idea that says more about our dietary relationship with the fungi. Cases of yellow knight poisoning reported from Lithuania in a 2016 study included a fifty-six-year-old man who had eaten generous helpings of the mushroom three times a day for a week.[18] Other comparably fanatical patients had consumed the fungus daily for a week or more, which raises the crucial issue of dosage and the adage attributed to Paracelsus, the Renaissance physician: *dosis sola venenum facit*, the dose makes the poison. The same rule applies to everything we eat. Potatoes, for example, contain the toxic alkaloid solanine, although one would have to eat a sack of tubers at a sitting to be poisoned by them.[19] What if yellow knight and every other mushroom contains toxins, and dosage is the deciding factor in the development of poisoning symptoms?

Rodents fed huge quantities of a variety of powdered mushrooms show the same spikes in creatine levels found in humans poisoned with yellow knights.[20] The mushrooms in these unpleasant studies included shiitake and chanterelles, and other indications of tissue damage were observed in mice fed masses of white button mushrooms. The doses imposed on the mice were staggering, equivalent to force-feeding a human fifty portobello mushrooms per day. The value of these experiments, which did not identify any toxins in the mushrooms, lies in their implication that any mushroom that is safe in moderation might be harmful in excess. Almost every species seems capable of upsetting the digestive system, so we need to avoid eating any mushroom by the bucket.[21] On the other hand, people who enjoy collecting and cooking a few fresh yellow knights from time to time should continue doing so. This does not mean, of course, that anyone should flavor their omelets with a little diced death cap.

THE PRETTIEST POISONOUS MUSHROOM

One of the prettiest mushrooms in all creation sits at the ruinous end of the toxicity scale. This fungus is the poison fire coral, whose scarlet fingers emerge from leaves littering the woodlands of Asia and Oceana and shed spores from their silky surface. Fire coral bears a passing resemblance to the medicinal mushroom reishi, which is why it is sometimes picked in error and steeped in hot water to make tea. This snafu replaces the mushroom of immortality, which is the nickname for reishi, with the mushroom of skin and hair loss, blood disorders, and brain damage.[22] The fire coral contains potent poisons called satratoxins, and a pea-sized piece of the fungus has proven lethal. Bizarre symptoms of acral skin peeling on the palms of the hands and soles of the feet, and alopecia, distinguish poisoning by fire coral from every other mushroom. Damage to the brain and bone marrow are also unique to the fire coral.

Corals or club fungi fruit as colorful spikes that range from fine wires that stand tall and straight, like dried spaghetti, to fatter stubs, and branched forms that resemble antlers, menorahs, and tiny painted cauliflowers.[23] A few species are popular edibles, including multiple species of chunky corals known as *escobetas*, or scrubbing brushes, in Mexico.[24] More than one thousand species of coral fungi have been described, and a good number cause nausea and intestinal turbulence. The following entry in a guidebook illustrates the difficulties in foraging for edible ones: "*Ramaria flava* is reported to be edible but of only moderate quality; however, it could easily be confused with *Ramaria formosa*, which is seriously poisonous, causing stomach pains and diarrhea if eaten. There is another reason why this coral fungus should not be collected in Britain: it is a very rare find."[25] Anyone who seeks coral fungi must be a very serious mycophile. These daredevils look at all mushrooms as potential edibles, evaporating the toxins from some of the dangerous species by cooking and reveling in nature with something akin to the attitude of big wave surfers or free climbers.

My reversal of their behavior—treating all mushrooms outside a grocery store with suspicion—will seem farcical to these experimental mycophagists, who may counter with "no one gets out of here alive," as Jim Morrison, of the Doors, and many others, have opined. They are modern McIlvainians, whose forebear, Captain Charles McIlvaine (1840–1909), tasted every mushroom he found to enrich the descriptions in his doorstopper of a book, *One Thousand American Fungi*, published in 1900.[26] Commenting on a group of mushrooms known for their laxative properties, this indomitable veteran of the Civil War wrote, "Wherever and however they grow, Hypholomas are safe. I have eaten them indiscriminately since 1881."

To go beyond the obviously noxious mushrooms tarred by millennia of poisonings, the McIlvaine method is the only way to widen the list of harmful and harmless fungi. With the nickname Old Ironguts, McIlvaine did not offer the most trustworthy advice on safety, but as the sample size of consumers has increased, printed and virtual mushroom guides offer a wealth of information for foragers committed to a little research.

WHY SOME MUSHROOMS ARE VERY POISONOUS

Larger animals expend more energy chewing and digesting fruit bodies than they get back in calories, which suggests that the chemical pathways that manufacture toxins in mushrooms probably evolved to repel insects and other tiny animals like nematode worms.[27] These smaller invertebrates have the opportunity to avoid the watery filaments that make up most of the fruit body and eat spores and other specialized cells in the mushroom that pack storage fats. This nutritional reward is enough to power the growth of insect grubs because we find some of the poisonous fruit bodies riddled with larvae. These insects are obviously impervious to the toxins.

When mushrooms decided to become poisonous, in an unconscious evolutionary sense, they opted to make chemicals that attacked the fundamental workings of their insect enemies as well as their unintended human targets. Making chemicals with this kind of generalized hostility toward life carries the occupational hazard of self-poisoning, but mushrooms have reengineered their proteins to be less sensitive to them.[28] They also benefit from operating as immobile platforms. Unlike a poisonous snake, which slides around with onboard venom sacs, the mature fruit body does not need to maintain an active metabolism and can poison the tissues of its stem and cap without any consequences. All the mushroom must do is stay upright and avoid damaging the cells on the gill surfaces until they finish producing the spores. The extreme toxicity of mushrooms like the death cap probably developed as a defense against their enemies as they gained resistance via a form of evolutionary arms race.

WITCHCRAFT AND WISDOM

In countries like Germany and Poland, where collecting wild mushrooms is an important part of the national culture, skills in fungal identification have been passed from generation to generation.[29] This basic mycological wisdom triumphed over long-standing superstitions about mushrooms among these mycophilic people and allowed them to avoid death caps while harvesting chanterelles. Poisonings still happened, of course, but there was no point blaming the local witch when it was obvious that someone had picked the wrong species. Outbreaks of fungal poisoning in medieval Europe that slew whole communities were a different matter. These epidemics were mystifying, because the toxin came from a tiny fungus that sprouted on rye and other cereals and was ground into the flour to make bread. Nobody demonstrated a solid link between this fungus and the plague that became known as ergotism in the seventeenth century, and supernatural explanations prospered in the knowledge vacuum of the Dark Ages.

Ergotism erupted after rainy summers, striking villagers with nausea, headaches, and vomiting; they lost the feeling in their fingers and toes, suffered convulsions and tetanus-like cramps, became unable to speak, and bit through their own tongues.[30] This was exactly what one would wish upon one's worst enemies. Maddened by their afflictions, some sufferers felt that insects were crawling beneath their skin, which is a symptom called formication, and others were consumed by feelings of intense heat. The burning sensation was described as *ignis sacer*, or holy fire, and seen as a punishment from God. Gangrene set into limbs starved of blood, fingers and toes turned black and broke off at the joints, and arms and legs were lost in the most extreme cases. These were the wages of sin.

Gangrenous symptoms predominated in some outbreaks, convulsions in others. The fungus, *Claviceps*, produces ergot alkaloids that interfere with the flow of nerve impulses and muscle contraction, wreaking havoc throughout the body. The ergot poisons are examples of mycotoxins, which is a term reserved for the toxins produced by molds rather than mushrooms.[31] In addition to narrowing blood vessels and making muscles seize, some mixtures of the ergot alkaloids produce behavioral changes and hallucinations. This diversity of disorders reflects the chemical virtuosity of the ergot fungus. It is a one-stop bioweapons center, whose real aim in life is to eat cereals without any competition from other microorganisms or from animals. After infecting the rye plant, the fungus takes possession of one or more of the flowers, growing in the place of healthy grains and protruding from the ears as firm black pellets, curved like bananas. These are the ergots, whose name comes from a word in Old French for a cock's spur, which is a perfect simile. The toxic alkaloids are mixed into the flour when the ergots are milled along with the healthy rye grains.

Ergotism was recognized in the ancient world, with a plausible mention on an Assyrian tablet, but the earliest undisputed descriptions come from the Christian era, when rye bread became a staple in Western Europe. Ergotism became known as St. Anthony's fire, after the relics of the eponymous desert father became associated with miraculous cures for the scourge. Spontaneous abortion was another of the shocking effects of ergotism, which led to the deliberate use of raw ergots to assist with labor and to induce abortions in unwanted pregnancies. Purified ergometrine, which is one of the alkaloids, was introduced as a drug to prevent hemorrhage after childbirth in the 1930s and played a significant role in the reduction of the maternal death rate.[32] Ergometrine is also used to induce the delivery of the placenta. Another alkaloid, ergotamine, is used to treat migraine headaches, and a variety of synthetic drugs based on the structures of the ergot chemicals have other uses in medicine. This is an example of healing with poisons in Western medicine, which was a central principle of Chinese medicine more than a thousand years ago.[33]

Lysergic acid is a precursor for the ergot alkaloids, forming a chemical skeleton that the fungus decorates and prunes to craft a whole family of toxins.[34] This was isolated from ergot and used to produce LSD by Albert Hoffman, the famous Swiss chemist who discovered its psychedelic properties through self-experimentation in the 1940s. Hoffman went on to isolate another fungal alkaloid from mushrooms rather than ergots. This was psilocybin, which stars in chapter 9.

Lysergic acid and closely related compounds in the fungus are presumed responsible for the frightening hallucinations experienced by victims of ergotism in the Middle Ages and must have added to the sense among the villagers that they were dealing with the occult. Divine punishment was a popular interpretation of these dreadful visitations, along with witchcraft and demonic possession. The psychedelic horror of ergotism is thought to have inspired the nightmarish figures in the triptych *The Temptation of Saint Anthony*, painted on hinged wooden panels by Hieronymus Bosch in 1501, or thereabouts.[35] Flying fish circle the sky, and a winged demon carries Saint Anthony to a ghastly forum occupied by a man with the face of a pig and other monstrosities. Anthony gestures toward Christ, who stands in a grotto separated from the anarchy, and symbols of ergotism include a severed foot and a village burning in the background.

Ergotism may have caused the outbreak of delirium and seizures among girls in colonial Massachusetts in the 1690s, but the court in the infamous Salem witch trials favored a supernatural source. Nineteen villagers were hung for witchcraft, and a twentieth martyr, a man, was crushed to death under a plank weighted with rocks in an attempt to obtain a confession.[36] Although ergotism may have been at the root of this debacle, other causes have been suggested, and the arousal of mass psychogenic illness or hysteria within a community of religious extremists seems likelier than the fungus.

MYCOTOXINS FROM MOLDS

The fearful poisonings caused by fungi have led to mycological explanations for other events in world history, including the Plague of Athens, the Black Death, the rise of Calvinism, and the Great Fear at the start of the French Revolution.[37] The evidence for these connections varies from weak to nonexistent, but the idea that we can apply modern science to decipher ancient mysteries is often more appealing than the truth that some historical riddles are unsolvable. Facts can be fitted to theories, and even when the original musings are dismissed by objective scholarship they persist in the popular imagination. The Athenian Plague, described by Thucydides in the *History of the Peloponnesian Wars*, is an interesting case study. Many of the symptoms of the plague described by Thucydides, including burning sensations, violent spasms, and the loss of fingers and toes, dovetail with ergotism, and this connection was first proposed by a German toxicologist in the nineteenth century. Later experts on ergotism pointed out that the fungus could not have caused the deaths of thousands of Athenians in 430 BC because they did not eat rye, but the tie-in to fungi was resurrected with the suggestion that the ancient Greeks could have been wasted by toxins produced by a different mold growing on stored wheat, which was a staple.[38] This keeps fungi on the list of possible causes of the Athenian Plague, although an infectious bacterium or virus is a safer bet.

Outbreaks of food poisonings are obviously limited to the direct consumption of the toxin, but they can resemble the impact of an infectious disease when a large enough population is compelled to eat a lot of the same thing under conditions of famine. The Athenians might have been forced to eat moldy grain during the interminable siege of their city by the Spartans. A more convincing case of mold poisoning played out in the region of Orenburg on the Russian steppes in the 1940s.[39] Severe food shortages during and after the Second World War led people to eat fallen grain. Lying under the winter snow, a species of *Fusarium* began to rot the unharvested crop, permeating the cereal with poisonous trichothecenes. The syndrome produced by consuming these chemicals is described as alimentary toxic aleukia, or ATA. ATA begins with nausea, vomiting, and diarrhea, like some mushroom poisonings, and is followed by widespread hemorrhaging of blood vessels that can progress to organ failure. The loss of blood flow to the skin results in infections by bacteria that accumulate in disfiguring facial wounds. Estimates suggest that one in ten people in the region were sickened with ATA; the number of deaths is unknown. This disaster was as diabolical as one of the historical outbreaks of ergotism, with the difference that there was no question about the cause: no witch trials, no need for self-flagellation, just suffering caused by a mold.

Ergotism and ATA are a distant threat now, but we absorb traces of mycotoxins from our food every day.[40] They are an unavoidable part of the diet, because molds contaminate the entire food chain from cereals to fruits, vegetables, dairy products, and meat. Even farmed fish carry traces of mycotoxins from the feeds used in aquaculture. Mycotoxins flow in our blood whatever we choose to eat, which is another illustration of the way that the threads of the human relationship with the fungi permeate every facet of our lives.[41]

Aflatoxins are the most widespread and troubling mycotoxins in the modern diet. They are made by species of *Aspergillus* and tuck themselves into our DNA, where they cause mutations that can increase the lifetime risk of liver cancer.[42] Aflatoxins also damage the gut and cramp the immune system, making us more susceptible to viral infections. At high doses they can cause birth defects, and later in life they are implicated in neurodegenerative diseases. This is a pretty damning rap sheet. Aflatoxins pose the greatest risks in developing countries, where hot and humid climates stimulate the growth of the molds on corn, peanuts, and other staple foods. Kenya is considered a hotspot for aflatoxin poisoning.[43]

In the United States, the USDA monitors the level of aflatoxins in peanuts, almonds, and pistachio nuts. Even with this surveillance, it is impossible to shield these commodities from mold growth, and the USDA permits low levels of aflatoxins in peanut butter and other foods. Public interest in this issue has forced the manufacturers of peanut butter to address the presence of traces of these natural toxins in the iconic American sandwich spread at a time when they are also dealing with more widespread concerns about peanut allergies in children. Interest in aflatoxins has supported a premium price for Valencia peanuts from New Mexico, where the dry climate reduces mold growth on the crop before and after harvesting. Cats and dogs are also affected by aflatoxins and are highly susceptible to poisoning because their diet is so uniform. This has led to regulations on mycotoxin levels in pet foods.

Most of our interactions with mycotoxins occur when we absorb these foodborne contaminants from the gut, but the lungs are a second sponge for toxins. Agricultural workers are most at risk if they inhale dust from contaminated grains. Cultures of lung cells treated with afla-toxins lose some of their ciliary strength, meaning that the hairs that normally move mucus in the lung beat more slowly when the cells absorb the toxins.[44] This adds to the problems of allergic illness in workers who inhale spores from moldy grain in silo operations without state-of-the-art air purifiers and dust masks (see chapter 3). Occupants of homes and buildings that become blackened with mold spores after flooding may also be exposed to mycotoxins, but the available evidence suggests that the dosages would be too low to cause any harm.[45]

"Blackened" is the operative word here because a mold named *Stachybotrys*, vilified as the "toxic black mold," became a public health pariah when it was linked to lung bleeding and infant deaths in Cleveland, Ohio, in the 1990s.[46] This fungus grows on the thick paper that covers the surface of drywall or gypsum board when it becomes soaked with water and produces a whole gamut of mycotoxins. It is a nasty piece of work, but its role in the Cleveland tragedy is unproven. The spores of the black mold are sticky and are not blown into the air very easily. This limits the number of spores that anyone is likely to inhale and means that the dose of the toxin carried on these particles is vanishingly small. Under conditions of massive mold growth, however, the jury is still out. The developing lungs of infants are likely to be very sensitive to mycotoxins, and this mold may be dangerous when it grows in dense patches on nursery walls. It has also been suggested that volatile chemicals that evaporate from the mold colonies may become toxic in confined spaces.[47] The funky smells of damp houses, wet clothing, and spoiled food are produced by some of these compounds. While we cannot detect most of these gases, dogs have been trained to sniff out areas of mold contamination in buildings by following these scents. Putting aside the potential risks of mycotoxins and volatile compounds in homes, fungal asthma is an incontrovertible menace to children, which means that excessive mold exposure is always a serious problem for public health.

Mycotoxins are the natural chemical weapons used by fungi in their internecine wars of mold against mold, driven by the ceaseless competition for food.[48] Millions of tons of spores circulate in the atmosphere,

drizzling every surface, softly, invisibly, always. As each spore germinates, its hyphal threads nudge against other tiny mycelia and the battles
commence with the exchange of poisons between combatants. Some
mold strains collapse, others neutralize the incoming mycotoxins and
counter with their own antifungal cocktails. This biological warfare has
been fought by the fungi across hundreds of millions of years.[49]

In addition to fighting other fungi, ergot has extended its repertoire
of specialized alkaloids to strike bacteria, nematode worms, and insects.
This wholesale violence is needed because the ergots would be consumed by these soil organisms when they fall to the ground at the end
of the growing season. The fungus must ward off these pests to get
through the winter before its spores can infect the next cereal crop in
the spring. Our vulnerability to poisoning by this fungus is a gruesome
by-product of the conflict between the fungus and its insect enemies
with whom we share the same types of nerve and muscle cells. The
human body is equipped with some impressive detoxification systems,
but the ergot alkaloids and many other mycotoxins have retained their
virulence against us.

We have enjoyed much greater success in handling the widespread
mycotoxin that is made by yeast to disable other microbes when they
begin to grow in the sweetness of grape must and beer wort. The physi-
ological adaptation of the body to yeast alcohol has been refined over
hundreds of thousands of years, and our cultural relationship with this
mycotoxin is part of the story of the rise of modern civilization.[50]
Through its effects on the nervous system, alcohol ranks as a potent
psychoactive drug, although it is rarely described in this manner. This
special designation belongs to a subset of fungal metabolites that assume control of the brain and whose action is considered in chapter 9.

9

Dreaming

USING MUSHROOMS TO
TREAT DEPRESSION

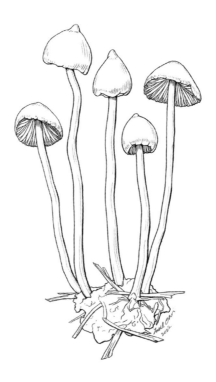

MAGIC MUSHROOMS light up the brain like fireflies in a meadow. Waves of nerve activity rise, crest, and dissolve from spot to spot across the brain, with islands of impulses crackling here, dampening there, as consciousness is disconnected from the usual flow of information. The brainwaves on mushrooms are similar to those in intense dreaming, with the twist that the temporary uncoupling from everyday thinking via the mushroom can have a lasting effect on our mindset when we reconnect. Anxiety and depression can lose some of their bite; life can seem less brutish. A mushroom dream is like a vacation to a tropical island or a canoeing trip along a pristine river, with the surprising benefit that the peace found during the break stays with you when it is over.

Exceptional dreams and dreamlike states that have been described as visions share many of the characteristics of a mushroom trip. In the Hebrew Bible, the prophet Ezekiel recalled a series of divine apparitions that he witnessed in Babylonia: "And I saw the creatures, and look, one wheel was on the ground by the creature on its four sides. The look of their wheels was like chrysolite [green gemstones], and a single likeness the faces of them had, and their look and their fashioning as when *a wheel is within a wheel.*"[1] I have been haunted by a similar vision since a luminous dream in which the night sky, powdered with stars, began to swirl, with pools of light revolving in the blackness. And as I looked, each wheel opened into more wheels, galaxies spinning inside galaxies and particles within atoms in a spectacle of infinite regress. For one glorious moment it seemed that nature was unveiling itself, the revelation increasing in magnification toward the germ of it all. Then I awoke, filled with wonder, and trying to hold on to the picture. The night sky has not danced in my unconscious since then. Like Caliban,

> The clouds methought would open and show riches
> Ready to drop upon me, that when I waked
> I cried to dream again

(SHAKESPEARE, *THE TEMPEST*, ACT 3, SCENE 2)

The details of the ancient prophecy and my dream were products of their time. Ezekiel was stirred by Mesopotamian imagery of fiery chariots and four-faced gods; my vision was drawn from reading popular books on cosmology, yet the sense of an epiphany seems similar. I cannot speak with authority for Ezekiel, but my fireflies came drug-free.

Magic mushrooms containing the psychedelic compound psilocybin provoke the same sense of transcendence over the commonplace perception of life. The brain on mushrooms produces some of the hallmarks of the rapid eye movement or REM stage of sleep but is heightened by the preservation of consciousness. The user is introduced to a form of lucid dreaming—dreaming while awake.[2] A straightforward mechanistic explanation of this process is elusive, which is not surprising in light of our bewilderment about how any of our emotions unfold in the nervous system. We know that love is made in the brain but have no notion of how it is encoded, accessed, augmented, or lost.

HOW PSILOCYBIN RATTLES THE BRAIN

After decades of scientific neglect, psilocybin has become a subject of intensive research, and some broad consensus is emerging on some of the neurological processes that govern the mushroom dream. Electrical impulses are conveyed along neurons through the movement of charged atoms or ions across their membranes. When these signals reach the end of the cells, they cause the release of chemical neurotransmitters that stimulate or block the generation of fresh impulses in the next neurons in the circuit. Serotonin is one of the neurotransmitters that perform this slower relay of sparks from cell to cell. When we consume psilocybin, a chemical group projecting from the ring structure of this little molecule is trimmed away in the liver, producing psilocin. The structure of the psilocin molecule is so similar to serotonin that it undoes the normal transmission of nerve impulses between cells.[3] Serotonin performs multiple roles in human physiology, ranging from the control of the unconscious process of digestion to the conscious emotion of happiness. If too much is released in the nervous system, the body responds with agitation and muscle cramping; too little and we lose motivation and can descend into depression.

To get a sense of the consequences of psilocybin consumption (psilocybin that is converted into psilocin), it is helpful to consider the complexity of the brain, which operates as an immense network of switches that turn on and off, relaying and blocking signals. Both ends of each nerve cell branch like fibrous tree roots to create as many as ten thousand connections between each of the hundred billion neurons in the brain, amplifying the signals through a circuit of one quadrillion (10^{15}) living transistors.[4] Comparisons with computers are fraught with difficulties, but the processing power of the brain matches a petascale supercomputer capable of one quadrillion computations per second.[5] With this complexity comes vulnerability, which explains the potency of the magic mushroom. Serotonin receptors are found throughout the nervous system but are particularly concentrated in the frontal cortex, which is the center of consciousness—our experience of life. As serotonin's twisted sister, psilocin activates some cortical circuits and stifles others, satirizing the world that we take for granted.

PHYSICAL AND EMOTIONAL SYMPTOMS OF PSILOCYBIN USE

Physical symptoms produced by the use of psilocybin result from the stimulation and repression of neural networks that normally respond to serotonin and include increases in heart rate and blood pressure, sweating, muscle twitching, facial numbness, nausea, lack of coordination, and headaches. These begin about twenty minutes after eating the mushrooms, differ greatly from person to person, and are usually mild. If we did not favor the uplifting psychological effects of psilocybin, these reactions would be seen as expressions of mushroom poisoning. Not on a par with the death cap, of course, but poisoning nevertheless, which is why many mushroom guides place a skull and crossbones symbol next to entries for *Psilocybe* species that contain the drug.

The psychological effects of psilocybin are similarly diverse. Some brain circuits are aroused and become overloaded with information while other parts of the brain are pushed into a dreamlike state. These changes in brain activity are visualized in patients who lie with their head inside the giant donut of a magnetic resonance imaging or MRI machine after they have consumed purified psilocybin. MRI experiments show cross-talk between parts of the brain that normally work in isolation, a reduction in blood flow to areas involved in logical thinking, and an increase in nervous activity in the deeper parts of the brain that control our emotions.[6] Sound and vision become tangled in some of these excursions, so that music is viewed as a kaleidoscope of colors. This de-segregation of the brain is called synesthesia. More frequently, our sense of individuality or ego dissolves, which leads to impressions of harmony and kinship with the rest of nature.[7] This is *henosis* and is the root of the encounters with God described by some psilocybin users.

Ego is lost when psilocin interferes with a brain circuit called the default mode network, or DMN. The DMN is concentrated in the prefrontal cortex and connects with hubs of neurons nestled farther back and deeper in the brain. Our sense of self is maintained in the DMN, and this is where the mushroom subverts our narcissistic programming. Imagine you are a passenger on a cruise ship that strikes an iceberg. In the seconds following the collision, normal activity in your DMN is suspended while other parts of the brain gather the information needed to figure out what has happened and plan a response. You are too frantic at that moment to be aware that you have raced onto the deck wrapped in a towel and wearing a shower cap. Ego has departed, albeit temporarily. Later, when it has become clear that the ship is sinking and that the lifeboats have left without you, the DMN has the last word as you resume contact with your sense of self, remove the shower cap, and are flooded with anxiety. Any feeling of positivity would be welcome in this hopeless situation, and this is where mushrooms can become our saviors. If you had swallowed a baggie of them when the ship's hull was ripped open, the psilocin would have disengaged the DMN from the alarming messages flowing from elsewhere in the brain, pushing you into a dreamlike state and leaving you more philosophical about the prospect of the frigid water.[8]

The soothing effect of magic mushrooms on passengers on a sinking ship is a matter of conjecture, but there is plenty of evidence that psilocybin can reduce our fearfulness in less dramatic situations. Multiple studies have shown that psilocybin is a useful treatment for clinical depression and can even foster a sense of well-being in patients with terminal illnesses.[9] In a trial conducted at Johns Hopkins University in 2016, patients with life-threatening cancer diagnoses reported feelings of greater life satisfaction after receiving high dosages of the drug. These improvements in attitude were sustained in 80 percent of the participants six months after their treatment. Follow-up interviews with surviving patients in 2020 revealed that the majority remained enthusiastic about their treatment, rating it personally meaningful and spiritually significant. Reflecting on the psilocybin treatment, one of the patients in the longer study wrote, "It's hard to explain. . . . Something in me softened, and I realized that everyone is just trying (mostly) to do the best they can. Even me. And that matters, since we are all connected." Another said, simply, "I have a greater appreciation and sense of gratitude for being alive."[10] These profoundly positive changes in outlook are astounding. Given the choice, how many of us would choose intimacy with psilocybin rather than raw cognition on the last lap?

Psilocybin has also proven useful in the treatment of PTSD and alcoholism, and researchers are studying whether the drug can be used to treat anorexia nervosa.[11] In addition to affecting the function of the existing circuitry in the brain, studies on mice show that psilocybin use can boost the number of connections between neurons.[12] The possibility, however slim, that a single dose can stimulate long-term rewiring of the brain has far-reaching implications for treating a host of health conditions. With depression and other mental illnesses afflicting hundreds of millions of people worldwide, psilocybin has already attained the status of a miracle drug in the opinion of some health-care professionals. With international attention to this unfolding research, pharmaceutical companies big and small are looking at magic mushrooms as a once-in-a-lifetime investment opportunity.[13] One of the challenges to this new industry is the long-standing illegality of growing mushrooms to produce psilocybin, but the mycological laws are evolving swiftly.

NORMALIZING PSILOCYBIN USE

In the U.S. elections in November 2020, Oregon voters passed Ballot Measure 109, legalizing the use of psilocybin for therapeutic purposes, by a comfortable margin. Measure 109 charged the Oregon Health Authority with developing the regulations for licensing growers to produce magic mushrooms and for licensed "facilitators" to provide "psilocybin services" in specialized clinics.[14] The authors of the measure pointed to the prevalence of mental illnesses in the region and crafted the legislation to "improve the physical, mental, and social well-being of all people in this state." Many health-care professionals objected; they wanted to see further studies and were concerned about the unspecified credentials of the facilitators who would counsel patients and shepherd them during their sessions. Unmoved by these arguments, the majority of voters looked forward to a happier future and scheduled the first clinics to open in 2023.

The most compelling case in favor of Measure 109 was made by people who said that their lives had been transformed by psilocybin. Mara McGraw was one of the brave proponents of the measure who shared her story with the media. Mara had undergone surgery, radiation, and chemotherapy for a rare form of neuroendocrine cancer for three years before she faced the necessity of making her end-of-life decisions. "After chemo failed, I went to a pretty dark place," she said.[15] Prescription antidepressants proved useless. Hearing about psilocybin treatment in Canada, she decided to give it a try. The psychoactive drug changed everything: "I felt an immediate release from the fear," she said in a video news conference. "I just felt fine and I felt like I rejoined everything in the universe." Her despair melted as the drug from the magic mushroom lit up her brain.

The evident benefits of psilocybin therapy outweigh the mild physical side effects of the drug, but there are concerns about the harmful psychological responses in a minority of patients. In a survey of almost two thousand patients, 84 percent said that they believed that they had benefited from psilocybin, which aligns with the results of many other studies.[16] Other respondents experienced a negative impact, with 3 percent saying that they had behaved aggressively or violently after consuming mushrooms and 8 percent seeking psychiatric treatment that they associated with the use of the drug. There have also been cases of self-harm and suicide attempts among psilocybin users, which speaks to the importance of taking the drug under some kind of supervision.

Some of the challenges to accepting psilocybin as a useful medicine are related to its popularity as a recreational drug, which began in the 1960s.

The negative image of psilocybin includes an enduring myth about an epidemic of magic mushroom and LSD users throwing themselves from buildings. Antipathy toward magic mushrooms is reinforced by the ludicrous claims of contemporary popularizers of mycology about the cosmic mysteries unveiled by psilocybin. Plenty of attention has been given to the historical players in the psilocybin story in other books, so a précis will suffice here: ethnomycologist Gordon Wasson and his wife, Valentina, excited millions of American magazine readers with their stories about the use of hallucinogenic mushrooms in Mexico in the 1950s.[17] Wasson collected hallucinogenic mushrooms in Mexico with Roger Heim, a French botanist, and Albert Hofmann, a Swiss chemist, who isolated psilocybin from cultures grown from these fruit bodies in 1958. Research on the effects of psilocybin was conducted by Timothy Leary at Harvard University in the 1960s, and Terence McKenna took "heroic doses" of psilocybin and published a guide to growing magic mushrooms in the 1970s.[18]

McKenna became a counterculture hero, and his increasingly deranged pronouncements discouraged more rational thinking about psychedelic mushrooms. One of his baseless claims centered on the chemical structure of psilocybin. McKenna declared that this was so unusual that it must have originated elsewhere in the galaxy. He went on to postulate that psilocybin mushrooms were a higher form of intelligence that had arrived from outer space and shaped the evolution of the human brain. Although he faces some stiff competition, McKenna's alien mushroom theory is one of the least enlightening things ever written about fungi. Returning to terrestrial mycology, recent genetic research has revealed a lot about the actual origins of magic mushrooms.

HOW MUSHROOMS MAKE PSYCHEDELICS

Psilocybin is produced by three hundred species of *Psilocybe* mushrooms and by other fungi belonging to three distantly related groups of fungi. *Psilocybe cubensis*, which grows on cow dung in nature, is cultivated indoors to furnish plentiful, year-round, and largely illegal sources of the drug. With interest in the therapeutic value of psilocybin overtaking its recreational use, researchers are investigating the industrial production of the drug in bacteria and yeast transformed with fungal genes.[19] The advantage of these microbes over the mushrooms is that they could produce huge quantities of pure psilocybin without any of the difficulties of raising and harvesting fruit bodies for drug extraction. This is a complicated project, with some similarities to the creation of the genetically modified (GM) microbes that produce insulin. If this is successful, psychedelic Frankenyeast will manufacture the drug in industrial fermenters in psilocybin breweries.

Psilocybin synthesis involves four enzymes that restructure the starting material, which is an amino acid called tryptophan, into psilocybin in the tissues of the mushroom.[20] The genes that encode these enzymes are clustered on a single chromosome in some of the fungi; in other species they are strung out and separated by genes that perform other functions.[21] Genetic comparisons between the species of magic mushrooms suggest that psilocybin synthesis evolved in one group of fungi and spread to other families by the process of horizontal gene transfer.[22] Most genes are transferred vertically, by inheritance from parent to offspring, but horizontal gene transfer between organisms in the same generation is common in some groups of microbes. It appears that psilocybin genes spread in this sideways manner from the mycelium of a magic species to the mycelium of an unmagical fungus and cast a spell on any animal that ate it. However this gene migration happened, psilocybin synthesis has prospered as a transferable skill in the fungi, which suggests that it must be doing something useful for them.

WHY MUSHROOMS MAKE PSYCHEDELICS

Insect attraction seems the best reason for a mushroom to make psychedelics, though the evidence for this is slim. Experiments suggest that flies, like humans, experience mood elevation on psilocybin. This is demonstrated by dropping flies in water and timing how long they continue to struggle to climb out.[23] Flies that have been fed psilocybin remain more active and keep trying to escape even when the situation seems hopeless. This seems a little like the patients in the human psilocybin study diagnosed with a terminal illness. Rather than providing a free antidepressant for the insects, the mushroom must profit from this interaction. Spore dispersal is the most important service provided for fungi by insects.[24] Flies called fungus gnats are hatched from eggs laid in psilocybes and could carry spores from their nurseries when they take to the wing as adults.[25]

This is all very speculative. However, the ability of a mushroom to attract flies by agitating their nervous systems aligns perfectly with the management of insect behavior by other kinds of microscopic fungi. Zombie ants, infected by tropical species of the fungus *Ophiocordyceps*, climb into the canopy of trees and bite down on leaf veins before the stalks of the fungus explode through their heads and spray spores into the air. This climbing behavior positions the fungus in the optimal location for the dispersal of its spores in wind. Other fungi as well as viruses and parasitic worms compel insects to behave in the same way. The uniformity of this response to infection suggests that the different parasites are exploiting the normal climbing behavior in insects for the purpose of dispersal.[26] Psilocybin has not been detected in these fungi, but it is produced by another fungus that infects cicadas and turns their abdominal tissues into masses of powdery spores.[27] Not content with sterilizing the insect, this parasite causes infected male cicadas to waggle their wings like females, attracting other males who receive a dose of spores when they mount their moldy mates. Compared with these astonishing feats of mind control, the mood-enhancing effect of psilocybin on insects and humans seems quite modest.

The colors of some mushrooms may be related to insect attraction, but this is an inconsistent trait among the hallucinogenic species. Psilocybes are inconspicuous brown mushrooms that seem unlikely to act as visual beacons for insects, although they do turn blue when they are bruised or broken.[28] The spotted caps of the fly agaric mushrooms offer a stronger visual cue that may lure insects. Pieces of these fruit bodies steeped in milk have been used as fly traps in Europe for centuries.[29] The polka dot pattern is destroyed when the mushroom is broken into the milk, so the fungus must also create a smell that attracts flies. Despite the use of the fungus as a fly trap, its name seems to refer to the devil, the lord of the flies, not to the insects.[30]

Fly agarics produce muscimol, which impersonates the neurotransmitter gamma-aminobutyric acid, or GABA, rather than serotonin. Neurons sensitive to muscimol have an inhibitory effect on the nervous system, reducing the transmission of impulses and acting as a sedative. There are more than a dozen different types of GABA receptor in the human brain, and muscimol binds more tightly to some than to others. This is the reason for the range of symptoms of fly agaric ingestion, including euphoria, lucid dreaming, changes in size perception, and feelings of weightlessness. These are described as the Alice in Wonderland syndrome, which is also recognized in patients suffering from viral infections, migraines, epilepsy, and brain damage.[31] The synesthesia provoked by psilocybin also comes with muscimol, but with its properties as a psychedelic "downer," this compound is an unlikely candidate for treating depression. (Other sedatives may be useful as antidepressants, including the synthetic drug ketamine, which binds to a different receptor than muscimol.[32]) Like psilocybes, fly agarics can cause serious poisonings in people who consume large doses of the mushroom or are unusually sensitive to the chemistry of this fungus.[33]

Human interactions are not part of the evolutionary programming of any of the magic mushrooms. We arrived too recently in their 150-million-year history to dictate the modification of the chemistry of these mushrooms by natural selection and, in any case, the huge numbers of prehistoric insects are bound to have been better at dispersing spores than humans and our ancestors. We get high on psilocybin and muscimol for the simple reason that we have the same brain chemistry as flies. Nevertheless, magic mushrooms may have exerted a supreme influence on civilization through religion, which is the most surprising extension of the human-fungus symbiosis.

MUSHROOMS AND THE CROSS

Familiarity with the contemporary human attraction to magic mushrooms, coupled with archaeological evidence and ethnographic studies, suggests that we have consumed fruit bodies containing psilocybin and other hallucinogenic compounds for millennia.[34] Rock carvings in

North Africa featuring bizarre figures holding and sprouting mush-rooms bolster ideas about the ritual use of fungi in the Neolithic, al-though we know nothing about the beliefs of the artists. The oldest clear evidence of mushroom worship comes from Mesoamerica and includes pre-Columbian stone carvings and pottery with fruit-body shapes, paintings of Mixtec gods offering mushrooms, and descriptions of the ceremonial consumption of mushrooms by Aztecs. Uninformed by neuroscience, our ancestors were bound to have interpreted the as-tounding effects of psychotropic fungi in supernatural terms. Minutes after receiving these mycological sacraments, their gods would have materialized in the form of dazzling creatures and mushroom men bear-ing prophesies from the sky. Ezekiel seems to have seen something like this following his vision of the psychedelic chariot, with "the look of fire with radiance all round. Like the look of a rainbow . . . the likeness of the glory of the Lord."[35] It is not surprising that there has been a good deal of speculation about Ezekiel's experimentation with mushrooms.

Author Robert Graves recognized Ezekiel's description in his own experience of an orchard paradise after eating psilocybes.[36] Graves had been in Mexico at the time with his friend Gordon Wasson. Beginning with the shamanistic use of fly agarics in Russia, which had been re-ported by European explorers in the nineteenth century, Wasson pur-sued evidence of the ritualistic use of mushrooms in other parts of the world. In his book, *Soma: Divine Mushroom of Immortality* (1968), Was-son identified the fly agaric as the vital ingredient in the ritual drink described in the Vedic Sanskrit text, the *Rigveda*.[37] A decade later he argued that LSD-like alkaloids from the ergot fungus stimulated the visions in ancient Greek ceremonies known as the Eleusinian Mysteries. But despite his enthusiasm for mycological explanations for religious practices, he was unmoved by the possible roots of Christianity in the worship of the fly agaric. This idea was developed by John Allegro, an English archeologist, who wrote *The Sacred Mushroom and the Cross* (1970).[38] Allegro was convinced that he had discovered a Sumerian code hidden in the Greek text of the gospels, which revealed the beliefs of a fertility cult whose rituals included the consumption of hallucino-genic mushrooms.

Allegro's bizarre translation of the Bible was rejected by myriad scholars, and the book was mocked in reviews. An eminent theologian writing in *The Times* of London spoke for many when he dismissed his psychedelic fertility cult as "a sensationalist lunatic theory."[39] Allegro's book included a photograph of a wall painting of Adam and Eve in Plaincourault Chapel in France, which features the serpent coiled around the Tree of Knowledge holding the forbidden fruit in its mouth toward Eve. Following the descriptions of earlier authors, Allegro believed that the thirteenth-century artist had pictured a giant mushroom tree bearing the spotty caps of fly agarics. Wasson favored the more conventional interpretation of the tree as a stylized Italian pine and rejected Allegro's translation of the Bible. This is interesting, given Wasson's credulity when it came to other ambiguous instances of mycological symbolism. It has been argued that financial ties with the Vatican through his banking career with J. P. Morgan & Co. may have swayed his opinion when it came to Christianity.

Whether or not the Plaincourault painting shows a mushroom or a pine tree, very credible mushrooms appear in other medieval artworks, including wall paintings in other churches in France and in Turkey, stained glass windows in Chartres Cathedral, a German tapestry, and the famous Great Canterbury Psalter.[40] Some of these compositions are mushroom-shaped objects rather than fungi, but an objective review of the surprising variety of images suggests that mushrooms were accepted as important religious symbols in the early church. Why else would they have been displayed in these commissions? Although his translations of scripture are fanciful, it seems that Allegro had recognized a genuine trace of mycology in Christianity.

Superstition and religion seem hard-baked into human nature. The scientific impulse is innate too: at our best we live by logic and conduct simple experiments when we have the opportunity to test our ideas. But in our age of scientific fruits and failures, the sky gods persist as invisible helpmates and moral judges across the planet. Faith in most gods would have developed without psilocybin and muscimol, but it is easy to imagine that their psychedelic power led to the ritual use of magic mushrooms and the invention of priesthoods. Without claiming that the Bible is a coded message from a fertility cult, perhaps mush-rooms did play a role in the origins of Judeo-Christian practices. Wall paintings in churches do not lead this inquiry very far, but they do show that mushrooms are intertwined with the ancestry of monotheism. If we had never swallowed the fruit bodies, human cultures might have evolved along very different agnostic trajectories.

We can take this mycological exploration of faith a step further to question the reality of any god. Mystical feelings of various kinds are a common response to psilocybin use in controlled experiments.[41] Two-thirds of the respondents in one study who said that they were atheists before using psilocybin changed their minds after consuming magic mushrooms and believed that they had encountered some form of "ultimate reality."[42] The faithful changed their minds too. Those who identified themselves as monotheists before taking psilocybin tended to lose this belief in favor of a broader idea of a benevolent intelligence in the universe. So rather than placing us closer to some holy spirit or demon who was waiting in the wings, it seems more logical to conclude from these studies that the alkaloid in these fungi conjured the gods via an avalanche of nerve impulses. There is a god-producer in the brain rather than a god-receptor.[43]

PSILOCYBIN AND THE ENTROPIC BRAIN

What do the effects of psilocybin say about the quotidian experience of life, that users feel so elevated and rank the effects among their most profound adventures? Aldous Huxley (1894–1963) was the most eloquent champion of the idea that psychedelic drugs overcome the "reducing valve" of the brain, which filters a surfeit of sensations gath-ered by our senses and feeds the conscious brain with slivers of informa-tion needed to stay alive.[44] In *The Doors of Perception* (1954), Huxley recounted his experiments with mescaline extracted from the peyote cactus. Mescaline is a serotonin agonist, like psilocybin and LSD, and produces similar psychedelic effects. High on the drug, Huxley looked at a flower arrangement and glimpsed "what Adam had seen on the morning of his creation—the miracle, moment by moment, of naked existence." Without psychedelic drugs, Huxley wrote that our awareness is limited to "the ruts of ordinary perception." In his earlier novel, *Brave New World* (1932), he had imagined a global palliative called soma, after the Vedic libation, which soothed and pacified the citizens of the World State.[45]

The opening of the reducing valve has been described as an increase in neural entropy.[46] This is entropy in a metaphorical sense rather than an authentic measurement of physical disorder. In the normal conscious state, we access information via a small subset of the available pathways. When mushrooms are added to the nervous system, the entropy of the brain increases as the connections between nerve cells spread across multiple regions, producing synesthesia, the loss of ego, and other expressions of enhanced networking. Obsessive-compulsive behavior is a good example of the kind of rigid thinking or low-entropy brain activity that can be relieved with psilocybin. Clinical depression is another. Caffeine has a comparably uplifting entropic effect. Coffee has no proven use in treating any mental illness, but the modest increase in connectivity and entropy probably explains the boost in creativity that is a blessing of the first cup in the morning—the fleeting sense of genius that is replaced by the lesser wit of the day. REM sleep has some similarities to this high-entropy state too. At the extremities of brain entropy, way beyond the normal effect of psilocybin, the psychotic brain illustrates the disastrous outcome of a surfeit of connectivity or disorder. McKenna may have found his way to these extremities of brain entropy with his heroic doses of the drug. To each their own.

In the search for relief from depression and anxiety, it is useful to consider why deep unhappiness is so prevalent. As imperfect products of evolution, there must be a natural imperative at work. Psychologists have wrestled with this question for decades, and although there is no completely satisfying answer, depression seems to emerge from a combination of bad wiring and the cryptic advantages of wariness, self-doubt, and sadness.[47] The bad wiring is strung between the more primitive lizard brain and the outermost cortex, where our consciousness and sense of self interact with the primal urges to feed, escape, attack, and copulate. Wariness is inevitable and protective, but the power of negative thinking is more problematic. Mild depression might be useful if it provides us with the opportunity to ruminate on a problem and reach a solution. It could also serve as an incentive device that is so unpleasant that it urges us to move on from a shattering experience. Deep relentless depression has no purpose. In his study of depression titled *The Anatomy of Melancholy*, Robert Burton (1577–1640) wrote, "What cannot be cured must be endured."[48] Mushrooms offer an alternative.

TO SHROOM OR NOT TO SHROOM

As calls to legalize psilocybin use grow, and corporations and governments seek to control the production of the drug, we need to be mindful of the implications of urging such abundant happiness in a frequently mordant mammal. Mortal illness and the approach of death can open Huxley's reducing valve without shrooms. Author Chris Paling described eating toast, his first solid food after six weeks on intravenous support in hospital: "I spread the butter. The aroma of it melting into the slightly charred bread is intoxicating. . . . I chew slowly. Another bite. Heaven."[49] Close to his death in 1994, British playwright Dennis Potter described the blossom on his plum tree: "I see it is the whitest, frothiest, blossomest blossom that there ever could be, and I can see it. Things are both more trivial than they ever were, and more important than they ever were, and the difference between the trivial and the important doesn't seem to matter. But the nowness of everything is absolutely wondrous."[50]

It is sad that mortal illness and the approach of death are needed to taste the toast and see the blossomest blossom, but the euphoria felt in extremis or on psilocybin is unsustainable, and Huxley's "ruts of ordinary perception" keep us alive. But by opening the valve a little, we light up the brain wiring for deeper thinking and creativity. The downside with using mushrooms to do so is that they can create an ungovernable flow of information.[51] The resulting feelings of oneness with the universe and empathy with the gods are entertaining, but they will not solve any mysteries or lead to meaningful insights about the cosmos. Remarkably, however, these festivities appear to soothe brains damaged by childhood abuse or battlefield trauma and promise to ease the universal disquiet at the end of life. We have not found anything else with these amazing properties.

And from the mushroom magic that opens and closes the doors of perception and has given us religion, we examine the global symbiosis with fungi that preserves life on the surface of the earth in the final chapter. Without these wider ecological interactions, there would be none of the more intimate expressions of the human-fungus symbiosis.

10

Recycling

THE GLOBAL MYCOBIOME

KEPLER 1649c is an Earth-sized planet in the constellation of Cygnus, three hundred light-years from our solar system. With its close orbit to a small star, climate models suggest it is quite Earth-like in temperature.[1] If Kepler 1649c is watery, it seems likely to harbor life, and if it accommodates anything more complicated than our bacteria, it is certain to be populated with fungi. Confidence in this untestable hypothesis is born from understanding the essence of fungi on Earth. Our fungi are the agents of entropy, transforming the energy captured in the web of life into the raw materials for the continuous regeneration of the biosphere. Without a group of organisms with these properties, the ecosystems of rocky worlds like Kepler 1649c would stall as planetary compost heaps of untappable energy. They must be there.

Mycology will evolve in its own way on this alien world, but some of its features are a given. Kepler's fungi will grow as filaments that permeate the solid refuse and multiply as yeasts on surfaces and suspended in fluids. The shapes of these cells will differ in detail from our species, but not in kind, because filamentous molds and budding yeasts are streamlined solutions to the challenges of growing in solids and liquids. Natural selection will craft them within the constraints of the environment on Kepler as it has on Earth.

There will be mushrooms too, which will cast spores into the autumnal breeze of Kepler's sunny hemisphere. Not fly agarics, of course, but stalked platforms of one kind and another that score the same goal of sending genes into the future. Beyond these bare necessities, Kepler's mycology will be determined by the characteristics of the rest of the planet's biology. If there are brains, the fungi will have found ways to lure some of their owners to help with spore dispersal and to dispel others from eating them. Supportive symbioses with the molds are as inevitable as damaging mycoses, and every Keplerian will carry a mycobiome. Mycology will be part of astrobiology across the cosmos— if not on Kepler 1649c, then elsewhere in the galaxy on other potentially habitable planets. There is no reason to think that there is anything like a human anywhere else, but there is going to be something like a fungus.

Back on Earth, the indelible nature of the human-fungus symbiosis ensures that our collaboration will continue to evolve, unconsciously and consciously, for better and worse, in health and in sickness. This chapter begins with the greatest extension of this interrelationship, which may be the most difficult to appreciate because it is so distanced from the body. This is the life support system that exists in the soil and in the plants that energize every food chain on land and oxygenate the atmosphere. We depend on botany, and plants depend on fungi.[2] Fungi support plants through their mycorrhizal associations, as endophytes that live inside their tissues, and by forming a protective shield of hyphae on leaf surfaces. On the flip side of their botanical roles, fungi decimate crops, exterminate forest trees, and compost the lot.[3] Fungi are the resurrection and the life.

SYMBIOSES WITH PLANTS

Attempts to categorize many species of fungi that interact with plants as faithful mutualists (advantage fungus/advantage host), commensals (advantage fungus/host unaffected), or parasites (advantage fungus/ host destroyed) are futile because they slide between these categories. The distinction between servant and slayer is blurred by fungi that are mutualists or commensals with some plants and parasites of others, and by mutualists that attack the tissues of their hosts when they are weakened by drought or old age.[4] In previous chapters we have seen how some of the fungi that we encounter all the time show a similar shift in behavior from harmless or even supportive interactions with the body to lethal pathogens when the immune system is compromised. Returning to the mutualisms that support plants, ectomycorrhizas between mushrooms and tree roots have become part of the general knowledge of ecology. They are introduced to children in elementary school, illustrated in posters and dioramas in nature centers, and enter informal conversations about the environment. Ectomycorrhizal fungi support trees by clothing their root tips with mycelia, radiating hyphae into the soil, and delivering water and dissolved minerals to their partners. There is nothing charitable about this: clamped to the roots, the fungus drains as much sugar from the plant as it allows. Research on these fungi has shown that their mycelia can create networks of filaments between the roots of adjacent trees and act as pathways for sharing resources within forests.[5] The importance of these interconnections for the trees remains controversial, however, and it is possible that the fungi are the chief beneficiaries of these underground webs.

Mushrooms growing beneath trees can seem remote from our concerns, but mycorrhizas are critical for forest health, and we benefit from the resulting carbon capture, oxygen production, and water purification. Supplies of lumber and other forest products are also reliant on these underground symbioses. Equally inconspicuous synergies with fungi exist through agriculture, where a different kind of mycorrhizal connection nourishes crop plants. These symbioses are called arbuscular mycorrhizas, and they are established with species in most plant families. Arbuscule, meaning a shrublike structure, is the term for the delicately branched connections that these fungi plug into root cells. This internal linkage or endomycorrhiza contrasts with the ectomycorrhizas that grow on the surface of roots and squeeze between their outermost cells.

Rice, corn, and wheat provide half of the calories consumed by humans, and the roots of all three of these staple plants are endowed with arbuscules. The fungi in these symbioses act as natural fertilizers by mopping up nitrogen, phosphorus, and potassium in the soil and sharing them with their plants.[6] When the same trio of elements are provided to the crops as NPK fertilizer, the natural mycorrhizas disappear.[7] They are lost because the fungi are unnecessary when the cereal roots are bathed with a supernatural abundance of these elements in the soil. Something similar happens to the healthy mycobiome in the digestive system when we poison the gut microbes with fast food.[8] Refined sugars and artificial sweeteners are quickly absorbed into the bloodstream, bypassing the community of bacteria and fungi needed to process complex carbohydrates (chapter 5). Fungicides are another problem for mycorrhizas. Sprayed on crops to control the rusts, smuts, blights, and blasts that plague monocultures, these chemicals trickle into the soil and kill the beneficial fungi before they can form mycorrhizas with the plants.[9] This is like the trickle-down effect of medicated shampoos that control dandruff and have the potential to squash some of the supportive fungi on the rest of the skin (chapter 2).

Using the same molecular genetic methods that have been developed for work on the human mycobiome, plant scientists have documented the rise and fall of populations of mycorrhizal fungi in croplands. The disturbance or dysbiosis resulting from intensive farming has not been recognized as a great loss to agribusiness because crop productivity keeps increasing through the mechanization of agriculture, chemical control of weeds and pests, and introduction of ever more vigorous cereal cultivars.[10] Nevertheless, agricultural practices that promote mycorrhizas are becoming popular. No-till farming avoids tearing the mycelia of mycorrhizal fungi apart before the seeds are planted, and soil enrichment with manure stimulates fungal growth.[11] The addition of spores as seed dressings is another strategy that provides seedlings with a preparatory mycobiome that can jumpstart the formation of mycorrhizas.[12] Dusting seeds with supportive fungi is reminiscent of the baptismal coating of newborn babies with yeasts from the mother (chapter 1). Through these neonatal get-togethers, the roots of germinating plants are transformed into mycorrhizas, and we begin our lifetimes as myco-humans.

THE NECROMYCOBIOME

Fungi collaborate with plants and animals throughout their lives and rot them after death. Decomposition by fungi returns nutrients to the soil and carbon dioxide to the atmosphere. Fresh roots and their fungi soak up the minerals released by decay, leaves absorb sunlight and CO_2, and the great wheel of the carbon cycle keeps turning. Fungi mingle with bacteria, insects, and worms in a fallen tree, each contributing to the process of decomposition in a distinctive fashion. Mycelia of mushrooms use the pressure in their filamentous hyphae to force their way into the wood and release enzymes that turn the trunk into powder and pulp. Bacteria crowd along the surface of the hyphae, fermenting the chemicals leaking from the fungi; beetles gouge galleries through the wood where yeasts blossom in the damp darkness, roundworms puncture the hyphae to feed on their juices, and fungi retaliate with toxins and sticky traps. All of this happens relentlessly, year after year, until the tree vanishes. Hardened brackets and hoofs of perennial fruit bodies on the surface of the rotting wood are joined by annual flushes of fleshy mushrooms as the external evidence of the internal decomposition. Spores from these fruitings are dispersed in pursuit of new sources of food, driving cycle upon cycle of life, death, and decay.

Fungi rot animals quite differently. Plants are made from sugars linked in chains to form cellulose and other polysaccharides that make up the dry weight of the plant. The fungi are the champions of releasing sugars from these materials. Animals are made from proteins and fats that are more susceptible to breakdown by bacteria, but yeasts grow in the slurry of the dead intestines and filamentous fungi set to work on the tougher tissues. Together with maggots that writhe in the froth and beetles that nibble at the sinews, the bacteria and fungi form the necrobiome that gathers at the postmortem banquet to dissolve the dead into the soil.[13]

The fungi of the necromycobiome change as the human corpse bloats, enters the phase of active decay, and becomes skeletonized. In the bloat stage the gut microbes destroy the digestive tissues, releasing gases that distend the cadaver and force "purge fluid" from the nose and mouth. This is when the greatest diversity of fungi is found in the body, including *Candida* yeasts and the familiar *Aspergillus, Mucor,* and *Penicillium* molds.[14] In the active stage of decomposition the diversity falls, and a mixture of specialized molds and yeasts works alongside the bacteria and maggots that liquefy the skin, muscles, and internal organs. Skeletonization leaves little food for the fungi apart from the hair and nails, which are digested by the species that cause ringworm in life.

The inevitability of our eventual decay is a fact of life that most of us would like to ignore. But we gain wisdom by understanding and embracing the part that we play in this grand terrestrial circus. The German philosopher Heidegger, among others, suggested that the affirmation of our own limited timeline allows us to transcend everyday experience and seek greater agency in life.[15] Some people find solace in this meditation, and burial suits impregnated with fungi have been marketed as biodegradable attire for enriching forest ecosystems after our demise.[16] This posthumous contribution to fertilizing the woods is a laudable ambition, and "green burials" of all kinds are a less poisonous exit plan than the use of embalming chemicals to keep the body looking cadaverous. Unfortunately, however, the advertised colonization of the burial suits with mycelia of oyster and shiitake mushrooms is not going to aid human decomposition because these are white-rot fungi that digest cellulose. In the unlikely event that oysters and shiitakes came across the body of a pirate in the wild (who missed his traditional burial at sea), they would remove all trace of his wooden leg, but little else.[17]

SPOILING ART AND RESTORING SOIL

Wooden legs and everything else that we saw, chisel, and pulp from trees are prone to decomposition by fungi. Air and moisture condemn cut wood to decay without the defenses against fungi provided in the living tree or chemical preservatives in cut lumber. The seeds of destruction are resting in the soil and drifting in the air, always ready to strike. The oldest surviving woodwork is the 12,500-year-old Shigir Idol discovered in a Russian peat bog in 1890. The lack of oxygen preserved the chiseled face and zigzag etchings of the five-meter-tall larch wood figurine, which is more than twice the age of Ötzi the iceman (see chapter 7).[18] Civilizations came and went as the Shigir Idol rested in the bog, and the fungi erased all trace of their carpentry beyond rings of postholes found at Woodhenge, near Stonehenge, and other Neolithic settlements in Europe.

Paintings are damaged by fungi too. Millennia before the early Mesolithic Siberians carved the Shigir Idol, artists decorated the walls of the Lascaux caves with pigments ground from local minerals. Within a few years of their discovery in 1940, the paintings showed signs of corrosion as the breath and sweat of thousands of visitors increased the humidity of the caves and acidified the damp rock. Electric lighting installed in the vaults caused a green alga to spread over the walls, along with patches of mold.[19] The Lascaux caves were closed to the public in 1963, but the microbiological damage has persisted. The problems are intensified by insects that disperse fungi in the caves, including a mold that blackens the walls and ceiling.[20]

Michelangelo had to remove mold spots from the damp lime that served as the canvas for his fresco in the Sistine Chapel, and medieval wall paintings in churches throughout Europe are threatened by fungal spoilage.[21] Fungal hostility toward our art and artifacts is relentless. Whatever we produce, they do their best to dissolve. Manuscripts and

books in library collections become moldy if the climate is not controlled, film and videotapes can be ruined by fungi, and faces in photographs become blurred by tiny mycelia growing on the gelatin. Only
digital images archived in clouds are safe. Fungi spot shoes, handbags,
and everything else made from leather. Spots of mildew on a favorite
jacket develop for the same reason that a fungus grows on our skin. Try
as we might, we cannot insulate ourselves from the fungi.

In his 1665 masterpiece *Micrographia*, Robert Hooke published the
earliest images of microscopic fungi, including a bread mold growing
on a sheepskin book cover. Centuries later we are still playing catch-up
with the universe of organisms and objects revealed with Hooke's microscope. Whether we see them or not, there is a fungus on everything,
decomposing its substance or sitting there as spores. Fungi have been
cleaning up the mess made by the rest of biology for hundreds of
millions of years, turning dead plants into compost and compost into
soil, threading their way through animal dung and, as we have seen,
dissolving the fibrous parts of animal corpses.

These skills in recycling are vital for soil regeneration after forest fires,
and mycorrhizas can help plants regain a roothold in land deforested by
timber harvesting and mining operations. We can also use mycelia to
break down many of the nastiest pollutants that we release into the environment and to reconfigure other chemicals to reduce their toxicity.[22]
White rot fungi use some of the enzymes that are effective in wood
decomposition to detoxify cancerous hydrocarbons produced when
fossil fuels are burned. They are good at this because the ringed structure of these molecules is similar to the lignin in wood that they are
accustomed to rotting. Other fungi are effective at breaking down agricultural pesticides and herbicides, pharmaceutical wastes, dyes, and
detergents. Mycelia also clean soils by concentrating toxic elements
from the water that trickles over their hyphae. Through this natural form
of filtration, fungi may even help remediate radioactive soil.[23] Although
we are a long way from extending this flair for detoxification from the
lab to the farm field and industrial site, pilot studies on these remarkable
processes offer a welcome distraction from the continuous newsfeed of
planetary gloom.

FASHIONABLE FUNGI

Highlights of this science have trickled into popular culture, where mushrooms have been embraced as the instruments of recycling that refresh the planet and support new life. This newfound love of mycology echoes the associations between mushrooms and fertility made by indigenous people across the world.[24] According to their traditional beliefs, the Blackfoot Indians imagined that giant puffballs, or *kakató'si*, were created by fallen stars. They painted the fruit bodies as white circles arising from a dark band along the bottom edge of tipi covers to symbolize the birth of life.[25] Now that the global scale of environmental damage is beyond any sensible question, the fungi have become widespread symbols of hope. After three hundred years of esoteric research and public disdain, fungi have become sexy.[26] Mushrooms have been embraced as emblems of beauty and countercultural cool in film and fashion, music, best-selling books, and inspirational lectures. Art installations with mycological themes have included giant mushrooms made from woven willow branches, living sculptures of heads grown from mycelia on wood chips and bristling with fruit bodies, and elaborate carvings and metalworks. Mushroom jewelry is very trendy too.

Ofer Grunwald, an Israeli artist, and his colleagues have created dot paintings with *Aspergillus* spores in tiny drops of agar jelly. Applied to sheets of glass, the drops form visible patterns when the spores germinate into tiny mycelia that color each dot.[27] Some of the designs are influenced by contemporary Australian Aboriginal art, and the participation of the fungi adds an extra dimension of individuality to every dot in the paintings. When the spherical spore of the mold germinates, its first thread can come from any point on its surface. The placement of the first branch to emerge from this hypha is similarly mutable, and the position of the second branch, and the branches from branches, so that within an hour of growth the tiny mycelia assume unique shapes in their drops. Although there is a high degree of predictability in the overall form of the growing fungus, its detailed geometry is a one-time creation. The colony is like a snowflake, whose precise details arise at one place and time in the universe and will never occur again. (This not as impressive as it sounds, perhaps, because nothing in biology is ever repeated. Even when cells and embryos have identical genes, they are unique in their physical minutiae.) Time-lapse photography captures the emergence of shape and color in the dot paintings over two or three days. There is a sense in which the arrow of time is reversed in this act of creation: rather than destroying works of art, the molds make art in Grunwald's hands by extracting energy from their jelly.

The creative impulse of the fungi is also expressed in vegan leather made from sheets of mycelia cultured in shallow trays and other fabrics produced by compressing blocks of mycelium grown on grains and wood chips. These materials have been crafted into handbags and clothing by famous designers and advertised as eco-alternatives to leather goods.[28] Vegan leather has also been adopted by shoemakers, which reverses the mildewing of shoes by fungi to the manufacturing of shoes by fungi.

QUEER MYCOLOGY

As mycology follows this new phase of its evolution, superstitions about the fungi continue to influence opinions about their unimportance on one side and their overwhelming significance on the other. This continuum of responses runs from mycophobes who dislike everything fungal to fanatics who believe that fungi can save the planet. In this vein, Patricia Kaishian and Hasmik Djoulakian have proposed that mycology is harmed by pervasive *mycophobia* that can be understood from the perspective of *queerphobia*: "Mycology is a science that, by its very nature, challenges paradigms and deconstructs norms. Mycology disrupts our mostly binary conception of plants versus animals. . . . Fungi are seen as poisonous, agents of disease, degenerate, deadly, freaky, gross, and weird—language historically leveled against both queer and disabled people—and as having no positive interrelationships with their environment(s)."[29] It is certainly true that fungi have suffered centuries of stereotyping that has burdened mycologists and inhibited progress in understanding their biology. Mycology has always been a nonconformist field. Kaishian and Djoulakian suggest that, although this has led many people to conclude that the fungi are "perverse and unworthy of formal investigation," others have found their strangeness inspiring. This tension has created tight-knit groups of researchers who work outside the better-known scientific disciplines, but has also encouraged frustrating ideas about the supernatural powers of fungi as medicinal cure-alls and environmental saviors. It can be difficult for the real science of mycology to overcome these half-truths and falsehoods.

Changing perceptions of the fungi are palpable among professional biologists. For most of the previous century, articles on biodiversity in scientific journals guesstimated the number of animal and plant species and skipped the fungi. Plant ecologists went about their business as if fungi did not exist, or virtue-signaled in seminars by mentioning mycorrhizas, and there seemed no place for mycology in zoology. But today, the fungi are on full display in pyramids of species, often as fly agaric icons; mycorrhizas are part of general biological knowledge; and the gut mycobiome of every animal is being scrutinized. This level of awareness seemed out of reach when I began my research career. Like the parents of actors concerned about their children's career choice, my dad was troubled when I told him that I intended to specialize in mycology for my doctoral degree—enough to consult a mycologist who happened, conveniently, to have retired in our Oxfordshire village. This was C. T. Ingold (1905–2010), a legendary figure in twentieth-century mycology who spent seventy years studying fungal spores.[30]

Ingold told dad that the study of fungi was an outstanding choice for a young scientist and that there would be dedicated departments of mycology in the universities before long. This was an overreach. The number of academic mycologists has actually declined since Ingold's forecast, and mycology departments are as scarce as hen's teeth.[31] On the other hand, researchers specializing in the study of medical mycology and plant diseases have attracted significant funding, and fungi are included in many areas of ecological research. And although they do not call themselves mycologists, yeast geneticists and biotechnologists who work with fungi are also employed in most research universities. As the classical taxonomists who named and organized the fungi have retired, mycology has emerged from the dust of their herbaria. Mycology has evolved from the study of isolated species to the interactions between fungi and other organisms.

THE CONSCIOUS MYCOBIOME

The study of these interactions with the human body is brimming with possibilities because the science has so far to go and the life of the mycobiome is new to our consciousness. Most of the time we have no awareness of the activities of the mycobiome at all. Our trillions of cells go about their business, and the fungi go about theirs. We may wonder why our scalp is itchy and what makes us sneeze when we brush past a mildewed houseplant, but the fungi do not distract us from the to-do list for the day. There is an equity between the involuntary reactions of the body to the presence of fungi and the behavior of the yeasts and molds as they feed on the waxes of the scalp, snuffle around in the gut lining, and fight the immune system.[32] We are complete coequals at the cellular level: fungal cells are every bit as perceptive and responsive as human cells.

The sensitivity of fungal cells is axiomatic. Hyphal filaments detect ridges on surfaces, grow around obstacles, and deploy a patch and repair system when they are injured. They react to confinement too, altering their growth rate, becoming narrower and branching less frequently. This allows them to adapt to the texture of the soil and the anatomy of plant and animal tissues as they push ahead and forage for food. Fungi also show evidence of learning and memory in experiments, responding to stress more effectively after training with a heat shock and growing in the direction where they found food in the past (see chapter 4). They are not thinking, in the sense that a brained animal thinks, but the fundamental mechanisms that allow a hypha to process information are the same as those at work in our bodies. Every thought in our lifetimes of thinking is processed by the billions of nerve cells in the brain and draws on cascades of reactions between proteins and other molecules. Every response of a fungal mycelium to its environment involves related cascades of signaling molecules. The difference between thinking and reacting is a matter of scale rather than essence. How could it be any other way, when all life is made from cells?

Estimates of the density of hyphae in grassland suggest that there can be between 10 billion and 1 trillion hyphae in one cubic meter of soil, and as many as 130 trillion hyphal tips cultivated in the same volume of straw or sawdust. These numbers are comparable to the density of neurons in the human brain, although nervous systems amplify their processing power by forming synapses that allow each nerve cell to connect with thousands of neighbors. Despite the incredible numbers of hyphae, the potential for communication is probably limited to the slow passage of chemical signals, and the fungus is unlikely to be relaying anything other than "I'm hungry," "I just found food," and "Will you mate with me?" This covers most of human discourse too, of course, but you get the point. Fungi do not dream or dread.

Claiming any consciousness for the fungi is dangerous territory. There is an eager audience for fantastical tales about fungi, especially if they are stretched from a few scientific observations. The inner language of mushrooms is a particularly awkward proposal, derived from the analysis of electrical spikes measured by sticking electrodes into blocks of mycelia.[33] The authors of this work concluded that the "complexity of fungal language is higher than that of human languages," based on a published scale in which French was identified as the least sophisticated European tongue. There is an event horizon in mycological thinking that is crossed when enthusiasm and ego exceed experimental evidence. One wonders whether potatoes employ a different syntax when their weak voltages are tapped to power LED clocks.

THE LUNAR EXTENSION OF THE MYCOBIOME

The interpretation of images of blobs on the surface of Mars from the Opportunity rover as puffballs and other mushrooms is a comparable flight of fancy.[34] This discovery came too late for Terence McKenna, who would have seen confirmation of his declarations about fungal spores arriving on Earth from space (see chapter 9). By the time McKenna developed his theory of cosmic migration, NASA had begun working in the other direction by transporting some of Earth's mycology *into* space. The Apollo missions extended the mycobiome to the Moon, when a dozen astronauts walked their fungi around on the lunar surface and left samples of the species from their guts along with urine and food wrappers in jettison bags. A photograph taken by Neil Armstrong during the Apollo 11 mission shows one of these white bags dropped from the lander. With lunar temperatures ranging from boiling to deep freezing, the bagged microbes are long dead, but their DNA will be readable if future visitors retrieve the Apollo refuse.[35]

Decades before the invention of the terms "microbiome" and "mycobiome," NASA scientists studied changes in the "fungal autoflora" of the astronauts on the Apollo 14 and Apollo 15 missions. Their interest in the microbiology of the body was way ahead of its time. The first methods for identifying bacteria and fungi from their DNA were not developed until the late 1970s, which limited the analysis of the astronaut mycobiome to identifying the fungi that could be grown from skin swabs and samples of astronaut feces. Filamentous fungi and yeasts were isolated from the swabs, and *Candida* yeasts were cultured from the fecal samples collected before, during, and after the missions.[36] The most interesting observation was the loss of fungi during the spaceflights and on the lunar surface. The number of species declined in space because the astronauts ate sterilized food, and there were no fungi in space that could reinforce their surviving mycobiomes.

NASA has always expressed concerns about contaminating the rest of the solar system with our microorganisms, but they cannot be dislodged from spacecraft. The problems begin with the assembly of space vehicles in cleanrooms.[37] Bacteria and fungi dodge the best efforts to sterilize the air supply and materials carried into these facilities. The microbiomes of the engineers are another source of fresh contaminants. The spores of *Aspergillus* and other molds found in the cleanrooms must have bypassed the air filters, and *Malassezia* and *Candida* yeasts discovered on surfaces probably originated on the workers. A rarified mixture of fungi persists on the International Space Station (ISS) in orbit, and some of them are expected to multiply in the absence of their earthbound competitors in this closed environment.[38] Experiments have demonstrated the tenacity of some fungi subjected to high doses of ultraviolet radiation, but nothing can survive on the outside of spacecraft exposed to the desiccating vacuum of space and the cosmic rays that obliterate DNA. There are no mushrooms on Mars.

The fungi growing on the skin and in the guts of astronauts on the ISS change during the mission, with a significant increase in the abundance of *Malassezia* yeasts on the skin. This seems to be related to changes in the amount of sebum produced by astronauts in space.[39] As the duration of space missions extends over years, the natural mixtures of microbes on the bodies of the astronauts are bound to disappear. Pathogens could prosper with the loss of species that normally keep them in check, and these anomalies are likely to be amplified by changes in the immune systems of the astronauts. Along with cosmic radiation, bone loss without gravity, muscle atrophy, and psychological stress, the disruption of our fungi is a factor that may make human space exploration impossible.[40] The good news is that we do not need rocket science to survive on Earth. We have everything we need biologically—other than the will to overcome the appalling selfishness of human nature. Appreciating the fungi is part of this terrestrial mission. This can begin with something as simple as looking at a mushroom—this beautiful oddity of nature—or inhaling the wondrous scent of a handful of rotting pine needles. There is so much beauty in this orgy of decomposition.

————

During our foray into the mycobiome we have encountered yeasts that live on the skin and a community of fungi roosting in the digestive system. These permanent residents of the human-fungus symbiosis are joined by spores that cause allergies and invasive pathogens that produce lethal infections. As we have seen, we can also think of the extensions of the mycobiome through our interactions with mushrooms that have fed and poisoned humans for millennia, and hallucinogenic species that have aroused deep superstitions and inspired religions. Other cultural relationships with fungi include brewing and baking with yeast and the use of filamentous fungi to ferment milk and make cheese. Rusts and smuts that ruin crops, and molds that spoil our harvests and homes, are part of the story too. These are examples of our partnership and competition with fungi throughout history and overlap with the modern engineering of yeasts and molds to produce life-saving drugs.

This is the story of the fungi near and far, which support the biosphere by forming mycorrhizas with plants, rotting the wastes of biology, enriching the soil, and purifying water. Life without fungi is impossible. There are as many of them living on the human body as there are stars in the Milky Way and, more importantly, they have a far greater influence on our lives than all but one of these galactic incinerators. They are everywhere and will outlive us by an eternity: *in myco speramus.*

Appendix

GHOST GUT FUNGI

MYCOBIOME RESEARCH is challenging, and a lot of misinformation has made its way into published studies. The problems begin with the indirect nature of the experiments. We cannot watch the fungi as they make their way along the gut, and many of them will not grow in culture dishes when they are collected from samples of feces. The only option for studying our onboard fungi is to identify them from their DNA signatures. From the resulting species lists we can infer things about the activities of the fungi in the gut because we know, for example, that one fungus digests fats and another consumes sugars, that this yeast can cope with low levels of oxygen, and its relative interferes with bacterial growth. At a time when identifying viruses using the polymerase chain reaction, or PCR, has become routine, analyzing fungi from their DNA might seem straightforward. But fungal genomes can be thousands of times bigger than viral genomes, which presents some technical difficulties, and there is another problem in separating the *real* gut fungi and signals from *ghost* gut fungi. Real gut fungi live in the gut or are ferried along in food; ghosts are cases of mistaken identification.

To make sense of the strings of As, Ts, Gs, and Cs amplified from fecal samples, investigators compare the sequences of fungi with reference DNA sequences associated with species catalog in online databases.

This is called sequence alignment. Ghost fungi arise from fake strings of As, Ts, Gs, and Cs produced by faults in the sequencing process and from weak matches that come from using DNA sequences that are too short for an accurate reading. Additional problems can be traced to faulty databases in which 10 percent or more of the archived DNA sequences are linked to the wrong species names.[1] For these reasons, the more reliable studies disregard the rarer matches to species whose sequences do not appear frequently in the samples and compare longer DNA sequences before concluding that a positive identification has been made.[2]

These technical errors are compounded by the lack of mycological training among the scientists scrutinizing the mycobiome. This has led to the publication of comical lists of species allegedly found in human feces that include toadstools that fruit from termite mounds, fungi that live inside eucalyptus trees, and tiny mushrooms that grow below the water in Argentinian lakes.[3] The names of these ghost fungi come from real fungi found in nature whose DNA sequences overlap a little—just enough to be confusing—with the DNA sequences of fungi amplified from the fecal samples. I came across a particularly daft example of this at a scientific conference when I spotted the name of a stinkhorn on a poster display about the infant mycobiome. The student presenting this work expressed no qualms when I suggested that this was an unlikely match. This stinkhorn, I explained, was a relative of a mushroom that looks like an erect penis and attracts carrion flies to its slime-covered tip. The presence of this fungus in the body was as improbable as finding a rhinoceros. He countered that he was simply reporting the data obtained from his computer search.[4]

Some of my colleagues may accuse me of duplicity here because I have criticized the work of taxonomists by writing about long-standing problems with the naming of fungal species.[5] They have a point. If we abandon the practice of giving names to fungi, who is going to recognize that something is wrong with identifying a termite fungus or a stinkhorn in a mycobiome study? Having conceded this, it is essential to remain skeptical that fungi dubbed with the same Latin name belong to the same species and behave in the same way. This is vital for clinical studies, because some versions or strains of the same "species" can have very different effects on the human body than others. The diversity of the fungi is at once enthralling and frustrating.

A list of the fungal species identified correctly in a sample of feces comprises long-term symbionts that spend their lives inside us and transients relayed with recent meals. Most mycobiome studies report the relative abundance of the different microbes. Pie charts are a good way to illustrate these results. When *Candida albicans* occupies a 70 percent wedge of the pie, this shows that seven of every ten strands of fungal DNA extracted from the sample came from this common yeast. Continuing with this hypothetical analysis, a 20 percent wedge might be occupied by *Malassezia restricta* (which also lives on the scalp), with the remaining 10 percent of the pie divided among a dozen rarer species. This pie chart shows that most of the mycobiome is populated by two fungi, which is interesting, but it does not tell us anything about the number of cells. In a depleted mycobiome with a total of a hundred million fungal cells, the relative abundance estimate in the example suggests that 70 million *Candida* cells are present in the gut; in a richer mycobiome overflowing with 40 billion cells, there could be around 30 billion cells of this omnipresent yeast. The pie chart showing relative abundance would look the same either way, and the missing information on actual numbers could have significant health implications. Until recently, many studies have ignored this inadequacy and simply reported relative abundance, but a growing number of investigators are taking the next step and measuring numbers using real-time PCR or qPCR, where the q stands for quantitative.

Traditional or old-school PCR measures the amount of DNA that has been amplified at the end of the PCR reactions. The *Candida* reading in the example shows that 70 percent of the DNA at the end came from this fungus. In qPCR, the increase in DNA is measured as the cycles of the PCR reactions proceed.[6] This means that the signal from *Candida* will increase swiftly in the early cycles because there are a lot of *Candida* cells containing *Candida* DNA at the beginning. The signals from the rare fungi build more slowly because there is so little starting material. In the qPCR method, the amplification of the DNA is monitored using fluorescent dyes that bind to the DNA of the target species. To relate the intensity of the fluorescence to the number of cells, the same qPCR reactions are run on samples containing known numbers of fungal cells that have been grown in culture.

Notes

CHAPTER ONE

1. Nicholas P. Money, *Fungi: A Very Short Introduction* (Oxford: Oxford University Press, 2016). The fungal kingdom and the animal kingdom have been married in one of ten supergroups of organisms, called the *opisthokonts* in modern biology. This uninspiring name refers to the arrangement of shared cell structures called cilia and should be replaced with a more evocative name: *mycozoans* would be better.

2. *Candida* is the Latin name of a genus of fungi that contains two hundred species of yeasts. It derives from *candidus*, meaning white, which is the color of the colonies of these yeasts dotted on a culture dish. *Candida* species have been found in Biscayne Bay, Florida; in deep-sea sediments beneath the turquoise waters of the Bahamas; in lakes and rivers in Brazil; and in grassland and agricultural soils. *Candida* grows on plants and inside the guts of insects, birds, and other animals. It is simpler to list the places where *Candida* is absent than to list its residences. The human mycobiome supports a half dozen species of *Candida*. *Candida albicans* is the dominant vaginal yeast and is the most frequent *Candida* species found in the gut and elsewhere in the body.

3. The interplay between fungi and bacteria in all ecosystems is a growing area of research: Aaron Robinson, Michal Babinski, Yan Xu, Julia Kelliher, Reid Longley, and Patrick Chain, "A Centralized Resource for Bacterial-Fungal Interactions Research," *Fungal Biology* 127, no. 5 (2023): 1005–1009.

4. Patrick M. Gillevet, Masoumeh Sikaroodi, and Albert P. Torzilli, "Analyzing Salt-Marsh Fungal Diversity: Comparing ARISA Fingerprinting with Clone Sequencing and Pyrosequencing," *Fungal Ecology* 2, no. 4 (2009): 160–167.

5. Maonon Vignassa, Jean-Christophe Melle, Frédéric Chiroleu, Christian Soria, Charléne Leneveu-Jenvrin, Sabine Schorr-Galindo, and Marc Chillet, "Pineapple Mycobiome Related to Fruitlet Core Rot Occurrence and the Influence of Fungal Species Dispersion Patterns," *Journal of Fungi* 7, no. 3 (2021): 175; Golam Rabbani, Danwei Huang, and Benjamin J. Wainwright, "The Mycobiome of *Pocillopora acuta* in Singapore," *Coral Reefs* (2021), https://doi.org/10.1007/s00338-021-02152-4; Luigimaria Borruso, Alice Checcucci, Valeria Torti, Federico Correa, Camillo Sandri, Daine Luise, Luciano Cavani, et al., "I Like the Way You Eat It: Lemur (*Indri indri*) Gut Mycobiome and Geophagy," *Microbial Ecology* 82 (2021): 215–223.

6. Ibrahim Hamad, Mamadou B. Keita, Martine Peeters, Eric Delaporte, Didier Raoult, and Fadi Bittar, "Pathogenic Eukaryotes in Gut Microbiota of Western Lowland Gorillas as Revealed by Molecular Survey," *Scientific Reports* 4 (2014): 6417; Alison E. Mann, Florent Mazel, Matthew A. Lemay, Evan Morien, Vincent Billy, Martin Kowalewski, Anthiny Di Fiore, et al., "Biodiversity of Protists and Nematodes in the Wild Nonhuman Primate Gut," *ISME Journal* 14, no. 2 (2020): 609–622; Ashok K. Sharma, Sam Davison, Barbora Pafčo, Jonathan B. Clayton, Jessica M. Rothman, Matthew R. McLennan, Marie Cibot, et al., "The Primate Gut Mycobiome-Bacteriome Interface Is Impacted by Environmental and Subsistence Factors," *NPJ Biofilms and Microbiomes* 8 (2022): 12.

7. James Cole, "Assessing the Calorific Significance of Episodes of Human Cannibalism in the Palaeolithic," *Scientific Reports* 7 (2017): 44707. The body of an adult male weighing 66 kilograms (146 pounds) contains an estimated 144,000 calories.

8. Ghee C. Lai, Tze G. Tan, and Norman Pavelka, "The Mammalian Mycobiome: A Complex System in a Dynamic Relationship with the Host," *WIREs Systems Biology and Medicine* 11, no. 1 (2019): e1438.

9. Lawrence A. David, Corinne F. Maurice, Rachel N. Carmody, David B. Gootenberg, Julie E. Button, Benjamin E. Wolfe, Alisha V. Ling, et al., "Diet Rapidly and Reproducibly Alters the Human Gut Microbiome," *Nature* 505, no. 7484 (2014): 559–563.

10. The number of bacteria in the gut microbiome comes from Ron Sender, Shai Fuchs, and Ron Milo, "Revised Estimates for the Number of Human and Bacteria Cells in the Body," *PLoS Biology* 14, no. 8 (2016): e1002533. Metagenomic analysis of fecal samples suggests that more than 99 percent of the DNA sequences come from bacteria, with the remaining sequences associated with archaea, viruses, and eukaryotes. Fungi are the most abundant of the eukaryotes in the gut, and we can come up with rough estimates of cell numbers from the relative abundance of sequences, which varies from 0.03 to 0.1 percent of the total, corresponding to 11 to 38 billion cells. This estimate is rounded to a maximum of 40 billion cells in the text. The mass, cumulative length, and surface area calculations for the cells are based on spherical bacteria and fungi with diameters of 1 µm and 4 µm, respectively. There is a good deal of wiggle room in these figures, but they serve as a useful order-of-magnitude guide to the scope of the mycobiome. The thousand-to-one ratio of bacteria to fungi in the microbiome (0.1 percent) appears in several studies, including the following review article: Tonya L. Ward, Dan Knights, and Cheryl A. Gale, "Infant Fungal Communities: Current Knowledge and Research Opportunities," *BMC Medicine* 15 (2017): 30. The lower published estimate of 0.03 percent for the fungal abundance in the gut microbiome comes from Stephen J. Ott, Tanja Kühbacher, Meike Musfeldt, Philip Rosenstiel, Stephan Hellmig, Ateequr Rehman, Oliver Drews, et al., "Fungi and Inflammatory Bowel Diseases: Alterations of Composition and Diversity," *Scandinavian Journal of Gastroenterology* 43, no. 7 (2008): 831–841. The gut surface area measurement was published in the same journal: Herbert F. Helander and Lars Fändriks, "Surface Area of the Digestive Tract—Revisited," *Scandinavian Journal of Gastroenterology* 49, no. 6 (2014): 681–689.

11. Indications that fungi play a relatively minor role in the gut microbiome come from research showing that the gut mycobiome is monopolized by species delivered in our food, including yeasts in bread, and should not be classified as true colonizers: Thomas A. Auchtung, Tatiana Y. Fofanova, Christopher J. Stewart, Andrea K. Nash, Matthew C. Wong, Jonathan R.

Gesell, Jennifer M. Auchtung, et al., "Investigating Colonization of the Healthy Adult Gastro-intestinal Tract by Fungi," *mSphere* 3, no. 2 (2018): e00092-18. Thomas Auchtung and colleagues also found that frequent teeth cleaning reduced the levels of *Candida albicans* in the gut. Presumably, teeth cleaning removes this yeast from the mouth before it is swallowed, whereas people who are strangers to the toothbrush are more likely to harbor higher levels of *Candida* in their digestive systems. An earlier study advised caution in interpreting metagenomic data on gut fungi because the techniques are so powerful that they identify species that are present in such low numbers that their biological effects must be negligible: Mallory J. Suhr and Heather E. Hallen-Adams, "The Human Gut Mycobiome: Pitfalls and Potentials—A Mycologist's Perspective," *Mycologia* 107, no. 6 (2015): 1057–1073.

12. Katarzyna B. Hooks and Maureen A. O'Malley, "Contrasting Strategies: Human Eukaryotic versus Bacterial Microbiome Research," *Journal of Eukaryotic Microbiology* 67, no. 2 (2020): 279–295.

13. World Health Organization, *WHO Fungal Priority Pathogens List to Guide Research, Development and Public Health Action* (Geneva: World Health Organization, 2022), https://www.who.int/publications/i/item/9789240060241.

14. Daniel B. DiGiulio, "Diversity of Microbes in Amniotic Fluid," *Seminars in Fetal and Neonatal Medicine* 17, no. 1 (2012): 2–11.

15. Kent A. Willis, John H. Purvis, Erin D. Myers, Michael M. Aziz, Ibrahim Karabayir, Charles K. Gomes, Brian M. Peters, et al., "Fungi Form Interkingdom Microbial Communities in the Primordial Human Gut That Develop with Gestational Age," *FASEB Journal* 33 (2019): 12825–12837; Linda Wampach, Anna Heintz-Buschart, Angela Hogan, Emilie E. L. Muller, Shaman Narayanasamy, Cedric C. Laczny, Luisa W. Hugerth, et al., "Colonization and Succession within the Human Gut Microbiome by Archaea, Bacteria, and Microeukaryotes during the First Year of Life," *Frontiers in Microbiology* 8 (2017): 738.

16. Matthew S. Payne and Sara Bayatibojakhi, "Exploring Preterm Birth as a Polymicrobial Disease: An Overview of the Uterine Microbiome," *Frontiers in Immunology* 5 (2014): 595. Some studies have raised concerns about the formation of biofilms of *Candida* on IUDs: Francieli Chassot, Melyssa F. N. Negri, Arthur E. Svidzinski, Lucélia Donatti, Rosane M. Peralta, Terezinha I. E. Svidszinski, and Marcia E. Consalro, "Can Intrauterine Contraceptive Devices Be a *Candida albicans* Reservoir?," *Contraception* 77, no. 5 (2008): 355–359. There is some evidence that the presence of an IUD throughout a pregnancy can boost the number of fungi in the amniotic fluid. In rare cases, amniocentesis can also introduce fungi and other microbes into the birth sac: Yohei Maki, Midori Fujisaki, Yuichiro Sato, and Hiroshi Sameshima, "*Candida* Chorioamnionitis Leads to Preterm Birth and Adverse Fetal-Neonatal Outcome," *Infectious Diseases in Obstetrics and Gynecology* 2017 (2017): 9060138.

17. Between the ages of one and six months, the average daily intake of breast milk is 750 milliliters. One milliliter of breast milk contains 350,000 fungal cells: Alba Boix-Amorós, Cecilia Martínez-Costa, Amparo Querol, Maria C. Collado, and Alex Mira, "Multiple Approaches Detect the Presence of Fungi in Human Breastmilk Samples from Healthy Mothers," *Scientific Reports* 7 (2017): 13016. This means that we gulp down more than two hundred million fungal cells per day in the first months of life. Similar numbers of bacteria were detected in an earlier analysis of breast milk samples: Alba Boix-Amorós, Maria C. Collado, and Alex Mira,

"Relationship between Milk Microbiota, Bacterial Load, Macronutrients, and Human Cells during Lactation," *Frontiers in Microbiology* 7 (2016): 492.

18. Lisa J. Funkhouser and Seth R. Bordenstein, "Mom Knows Best: The Universality of Maternal Microbial Transmission," *PLoS Biology* 11, no. 8 (2013): e1001631.

19. Michael Obladen, "Thrush—Nightmare of the Foundling Hospitals," *Neonatology* 101, no. 3 (2012): 159–165; Thomas J. Walsh, Aspasia Katragkou, Tempe Chen, Christine M. Salvatore, and Emmanuel Roilides, "Invasive Candidiasis in Infants and Children: Recent Advances in Epidemiology, Diagnosis, and Treatment," *Journal of Fungi* 5, no. 1 (2019): 11.

20. "Caesarean Section Rates Continue to Rise, amid Growing Inequalities in Access," World Health Organization, June 16, 2021, https://www.who.int/news/item/16-06-2021-caesarean -section-rates-continue-to-rise-amid-growing-inequalities-in-access-who. Live births by C-section vary, from less than 20 percent in Israel and Scandinavian countries to 45 percent in South Korea and more than half of all births in Turkey.

21. "Infant and Young Child Feeding," UNICEF, last updated December 2022, https://data .unicef.org/topic/nutrition/infant-and-young-child-feeding/#; "Breastfeeding," Centers for Disease Control and Prevention, accessed July 25, 2023, https://www.cdc.gov/breastfeeding /index.htm. There is a lot of variation in the rate of breastfeeding across the United States, with more than two-thirds of babies in some states being breastfed for at least the first six months, declining to less than 40 percent of infants in Mississippi and Alabama.

22. Thomas A. Auchtung, Christopher J. Stewart, Daniel P. Smith, Eric W. Triplett, Daniel Agardh, William A. Hagopian, Anette G. Ziegler, et al., "Temporal Changes in Gastrointestinal Fungi and the Risk of Autoimmunity during Early Childhood: The TEDDY Study," *Nature Communications* 13 (2022): 3151.

23. Lene Lange, Yuhong Huang, and Peter K. Busk, "Microbial Decomposition of Keratin in Nature—A New Hypothesis of Industrial Relevance," *Applied Microbiology and Biotechnology* 100, no. 5 (2016): 2083–2096; Hermann Piepenbrink, "Two Examples of Biogenous Dead Bone Decomposition and Their Consequences for Taphonomic Interpretation," *Journal of Archaeological Science* 13, no. 5 (1986): 417–430.

CHAPTER TWO

1. Katarzyna Polak-Witka, Lidia Rudnicka, Ulrike Blume-Peytavi, and Annika Vogt, "The Role of the Microbiome in Scalp Hair Follicle Biology and Disease," *Experimental Dermatology* 29, no. 3 (2020): 286–294; Dong H. Park, Joo W. Kim, Hi-Joon Park, and Dae-Hyan Hahm, "Comparative Analysis of the Microbiome across the Gut-Skin Axis in Atopic Dermatitis," *International Journal of Molecular Sciences* 22 (2021): 4228.

2. This thought experiment begins with a size comparison of yeasts and humans. A yeast cell with a diameter of 4×10^{-6} m ($4\ \mu m$) has a cross-sectional area of 1.3×10^{-11} square meters (m^2). The floorspace occupied by a standing human with a modest allowance for arm movement is $0.1\ m^2$, which is thirteen billion times larger than the outline of a yeast. One million yeasts growing in a one square centimeter patch of skin fill 13 percent of the available space. Crowded like yeasts, the current human population would occupy $1/0.13 \times 8 \times 10^9 \times 0.1\ m^2 = 6.2 \times 10^9\ m^2 = 6,200$ square kilometers, which equals the contiguous urbanized area of Los

Angeles. This density represents a five-hundred-fold increase in the current population of Los Angeles.

3. Robert L. Gallo, "Human Skin Is the Largest Epithelial Surface for Interaction with Microbes," *Journal of Investigative Dermatology* 137, no. 6 (2017): 1213–1214. Gallo cites the widely accepted surface area estimates of 2 square meters (m^2) for the skin, 30 m^2 for the gut, and 50 m^2 for the lungs. If we include the invaginations of the hair follicles, sweat glands, and sebaceous glands, the epithelial surface of the skin increases to at least 30 m^2. A 140-by-70-centimeter bath towel has an area of 1 m^2.

4. The best estimates suggest that fewer than one hundred billion bacterial and fungal cells live on the skin, which compares with the estimated forty trillion occupants of the gut microbiome.

5. The highest levels of oxygen are found close to the wall of the gut, which is supplied by a rich system of blood vessels. Most of this oxygen is consumed by the microbiome and independent chemical reactions that keep the gut lumen anoxic: Elliot S. Friedman, Kyle Bittinger, Tatiana V. Esipova, Likai Hou, Lillian Chau, Jack Jiang, Clementina Mesaros, et al., "Microbes vs. Chemistry in the Origin of the Anaerobic Gut Lumen," *Proceedings of the National Academy of Sciences USA* 115, no. 16 (2018): 4170–4175. Some bacteria can live with or without oxygen. They are called facultative anaerobes. Very few fungi have this flexibility, which means that fungal growth must be limited to locations next to the interior of the gut wall.

6. Hye K. Keum, Hanbyul Kim, Hye-Jin Kim, Taehun Park, Seoyung Kim, Susun An, and Woo J. Sul, "Structures of the Skin Microbiome and Mycobiome Depending on Skin Sensitivity," *Microorganisms* 8, no. 7 (2020): 1032.

7. Zuzana Stehlikova, Martin Kostovcik, Klara Kostovcikova, Miloslav Kverka, Katernia Juzlova, Filip Rob, Jana Hercogova, et al., "Dysbiosis of Skin Microbiota in Psoriatic Patients: Co-occurrence of Fungal and Bacterial Communities," *Frontiers in Microbiology* 10 (2019): 438. Settled and well-defined communities of fungi are destabilized in cases of sensitive skin syndrome and psoriasis and replaced with different collections of fungi on each patient. This is a mycological instance of the Anna Karenina principle, or AKP—namely, all happy mycobiomes are alike, but each unhappy mycobiome is unhappy after its own fashion. The AKP has been applied to science, politics, and economics, wherever it seems that there are more ways for the subject that is being examined to be unstable and dysfunctional than to be stable and functional. Microbiologists have found that about half of all diseases associated with changes to the communities of microbes on the body follow the AKP: Jesse R. Zaneveld, Ryan McMinds, and Rebecca Vega Thurber, "Stress and Stability: Applying the Anna Karenina Principle to Animal Microbiomes," *Nature Microbiology* 2 (2017): 17121; Zhanshan S. Ma, "Testing the Anna Karenina Principle in Human Microbiome-Associated Diseases," *iScience* 23, no. 4 (2020): 101007. The reason that variety rules the mycobiome in some illnesses and a single fungus emerges in others may come down to the role played by the fungi. According to this idea, the Anna Karenina principle applies when fungi respond to an illness rather than causing it to develop, and multiple species flare up as the tissue damage unfolds. See discussion of colon cancer in chapter 5.

8. Geoffrey C. Ainsworth, *Introduction to the History of Medical and Veterinary Mycology* (Cambridge: Cambridge University Press, 1976).

9. Keith Liddell, "Skin Disease in Antiquity," *Clinical Medicine* 6, no. 1 (2006): 81–86.

10. The translation of Suetonius, *The Twelve Caesars*, by Anthony S. Kline, explains that Augustus used the scraper very vigorously to relieve his itching. The standard translations convey the false impression that the use of the scraper caused the skin blemishes, https://www.poetryintranslation.com/PITBR/Latin/Suethome.php. The quote about Festus comes from the poet John Donne, who wrote a defense of suicide in 1608: *Biathanatos*, ed. M. Rudnick and M. Pabst Battin (New York: Garland, 1982), 66. The classical source for this story was the Roman poet Martial, who did not specify that Festus was suffering from ringworm: "o'er his very face crept black contagion." This quote comes from Martial, *Epigrams*, vol. 1, ed. and trans. David R. Shackleton Bailey, Loeb Classical Library (Cambridge, MA: Harvard University Press, 1993), epigram 78, pp. 78–79. Donne's sources are evaluated by Don C. Allen, "Donne's Suicides," *MLN* 56, no. 2 (1941): 129–133.

11. John Aubrey, *The Natural History of Wiltshire: Written between 1656 and 1691*, ed. J. Britton (London: J. B. Nichols, 1847), 37.

12. Ainsworth, *Introduction*, 4–5; Richard Owen, "On the Anatomy of the Flamingo (*Phaenicopteris ruber*, L.)," *Proceedings of the Zoological Society of London* 2 (1832): 141–145. The bird dissected by Owen had suffered from aspergillosis caused by a species of *Aspergillus*. The earliest report of human aspergillosis involved a sinus infection in a French soldier in the eighteenth century: M. Plaignaud, "Observation sur un Fongus du Sinus Maxillaire," *Journal de Chirurgie (Paris)* (1791): 111–116. After several surgeries, the patient was cured with the use of a "branding iron introduced through the nose by means of a cannula. . . . The fungal growths, burnt to their root, never reappeared." The fungus that caused pulmonary aspergillosis was described by John Hughes Bennett, a British physician working in Edinburgh, who examined sputum samples from infected patients: John H. Bennett, "XVII. On the Parasitic Vegetable Structures Found Growing in Living Animals," *Transactions of the Royal Society of Edinburgh* 15, no. 2 (1844): 277–294. Under the microscope, Bennett saw "the most beautiful and regular vegetable structure" of transparent tubes with "joints composed of distinct partitions . . . constricted like certain kinds of bamboo." He also described "bead-like rows" of spores in the clinical samples. Infectious hyphae had been described in the previous century by William Arderon, who illustrated a freshwater roach whose tail was bristling with filaments: Ainsworth, *Introduction*, 3–4. This fish infection is caused by a microorganism classified as a water mold, rather than a fungus, and is known as saprolegniasis.

13. Editorial, "Robert Remak (1815–1865)," *Journal of the American Medical Association* 200, no. 6 (1967): 550–551; Andrzej Grzybowski and Krzysztif Pietrzak, "Robert Remak (1815–1865): Discoverer of the Fungal Character of Dermatophytoses," *Clinical Dermatology* 31, no. 6 (2013): 802–805. Other pioneers in the study of fungal infections of the skin included Johannes Lukas Schönlein (1793–1864) and David Gruby (1810–1898). Schönlein was inspired by the work of Agostino Bassi (1773–1856), who demonstrated that a fungus caused a disease of silkworms in the 1830s. Bassi was the first scientist to show that a microorganism could cause a disease in an animal.

14. Brian P. Hanley, William Bains, and George Church, "Review of Scientific Self-Experimentation: Ethics History, Regulation, Scenarios, and Views among Ethics Committees and Prominent Scientists," *Rejuvenation Research* 22, no. 1 (2019): 31–42. Experiments on gonorrhea and syphilis were performed in the eighteenth century by a British surgeon, John Hunter. Hunter may have inoculated one or more of his patients with infected pus rather than himself,

which would have been criminal as well as unethical: George Qvist, "John Hunter's Alleged Syphilis," *Annals of the Royal College of Surgeons of England* 59, no. 3 (1977): 206–209.

15. Most ringworm infections in humans are caused by species of *Trichophyton*. These are classified in a family of ascomycete fungi called the Arthrodermataceae. *Trichophyton rubrum* is the commonest cause of *tinea corporis*. *Trichophyton violaceum* is a very close relative that causes hair and scalp infections. Other species include *Trichophyton mentagrophytes*, which infects humans when it is transferred from dogs, cats, and other pets. Skin infections are also caused by species of *Epidermophyton*, *Microsporum*, and *Nanizzia*, which belong to the same family as *Trichophyton*. Readers interested in exploring the taxonomy of these fungi should consult the following sources: G. Sybren de Hoog, Karoline Dukik, Michel Monod, Ann Packeu, Dirk Stubbe, Marijke Hendrickx, Christiane Kupsch, et al., "Toward a Novel Multilocus Phylogenetic Taxonomy for the Dermatophytes," *Mycopathologia* 182, nos. 1–2 (2017): 5–31; P. Zhan, K. Dukik, D. Li, J. Sun, J. B. Stielow, B. Gerrits van den Ende, B. Brankovics, et al., "Phylogeny of Dermatophytes with Genomic Character Evaluation of Clinically Distinct *Trichophyton rubrum* and T. *violaceum*," *Studies in Mycology* 89 (2018): 153–175.

16. Brian B. Adams, "Tinea Corporis Gladiatorum," *Journal of the American Academy of Dermatology* 47, no. 2 (2002): 286–290; D. M. Poisson, D. Rousseau, D. Defo, and E. Estève, "Outbreak of Tinea Corporis Gladiatorum, a Fungal Skin Infection Due to *Trichophyton tonsurans*, in a French High Level Judo Team," *Eurosurveillance* 10, no. 9 (2005): 562.

17. Felix Bongomin, Sara Gago, Rita O. Oladele, and David W. Denning, "Global and Multi-National Prevalence of Fungal Diseases—Estimate Precision," *Journal of Fungi* 3 (2017): 57.

18. J. N. Moto, J. M. Maingi, and A. K. Nyamache, "Prevalence of Tinea Capitis in School Going Children from Mathare, Informal Settlement in Nairobi, Kenya," *BMC Research Notes* 8 (2015): 274.

19. Josephine Dogo, Seniyat L. Afegbua, and Edward C. Dung, "Prevalence of Tinea Capitis Among School Children in Nok Community of Kaduna State, Nigeria," *Journal of Pathogens* (2016): 9601717.

20. A. K. Gupta and R. C. Summerbell, "Tinea Capitis," *Medical Mycology* 38, no. 4 (2000): 255–287.

21. Morris Gleich, "Thallium Acetate Poisoning in the Treatment of Ringworm of the Scalp: Report of Two Cases," *JAMA* 97, no. 12 (1931): 851. In his paper, Gleich referred to the deaths of fourteen children in an orphanage in Grenada, Spain, who received an accidental overdose of thallium acetate for ringworm in 1930. A year after the publication of Gleich's paper, a British dermatologist endorsed the continued use of rat poison for treating ringworm: John T. Ingram, "Thallium Acetate in the Treatment of Ringworm of the Scalp," *British Medical Journal* 1, no. 3704 (1932): 8–10. Ingram wrote, "There is no serious evidence against the use of thallium acetate . . . [t]hough toxic symptoms may occasionally be encountered, they are seldom severe, and the patient invariably recovers," which was not very reassuring.

22. Anonymous, "'X' Rays as a Depilatory," *The Lancet* 147, no. 3793 (1896): 1296.

23. S. Cochrane Shanks, "Thallium Treatment of Ringworm," *British Medical Journal* 1 (1932): 121.

24. Rebecca Herzig, "The Matter of Race in Histories of American Technology," in *Technology and the African-American Experience: Needs and Opportunities for Study*, ed. Bruce Sinclair (Cambridge, MA: MIT Press, 2004), 179–180.

25. Roy E. Shore, Miriam Moseson, Naomi Harley, and Bernard S. Pasternack, "Tumors and Other Diseases Following Childhood X-Ray Treatment for Ringworm of the Scalp (*Tinea capitis*)," *Health Physics* 85, no. 4 (2003): 404–408.

26. Liat Hoffer, Shifra Shvarts, and Dorit Segal-Engelchin, "Hair Loss Due to Scalp Ringworm Irradiation in Childhood: Health and Psychosocial Risks for Women," *Israel Journal of Health Policy Research* 9 (2020): 34.

27. Esther Segal and Daniel Elad, "Human and Zoonotic Dermatophytoses: Epidemiological Aspects," *Frontiers in Microbiology* 12 (2021): 713532. Geophilic mycoses are caused by fungi that come from an external source in the environment like soil or decomposing plant material.

28. Andriana M. Celis Ramírez, Adolfo Amézquita, Juliana E. C. Cardona Jaramillo, Luisa F. Matiz-Cerón, Juan S. Andrade-Martínez, Sergio Triana, Maria J. Mantilla, et al., "Analysis of *Malassezia* Lipidome Disclosed Differences among the Species and Reveals Presence of Unusual Yeast Lipids," *Frontiers in Cellular and Infection Microbiology* 10 (2020): 338. Parasitic wasps that lay their eggs on caterpillars have followed the same evolutionary path as *Malassezia* and extract their fatty acids from their hosts.

29. Minji Park, Yong-Joon Cho, Yang W. Lee, and Won H. Jung, "Understanding the Mechanism of Action of the Anti-Dandruff Agent Zinc Pyrithione against *Malassezia restricta*," *Scientific Reports* 8 (2018): 12086.

30. Hee K. Park, Myung-Ho Ha, Sang-Gue Park, Myeung N. Kim, Beom J. Kim, and W. Kim, "Characterization of the Fungal Microbiota (Mycobiome) in Healthy and Dandruff-Afflicted Human Scalps," *PLoS ONE* 7, no. 2 (2012): e32847.

31. Diana M. Proctor, Thelma Dangana, D. Joseph Sexton, Christine Fukuda, Rachel D. Yelin, Mary Stanley, Pamela B. Bell, et al., "Integrated Genomic, Epidemiologic Investigation of *Candida auris* Skin Colonization in a Skilled Nursing Facility," *Nature Medicine* 27 (2021): 1401–1409.

32. Suhail Ahmad and Wadha Alfouzan, "*Candida auris*: Epidemiology, Diagnosis, Pathogenesis, Antifungal Susceptibility, and Infection Control Measures to Combat the Spread of Infections in Healthcare Facilities," *Microorganisms* 9 (2021): 807.

33. Nancy A. Chow, José F. Muñoz, Lalitha Gade, Elizabeth L. Berkow, Xiao Li, Rory M. Welsh, Kaitlin Forsberg, et al., "Tracing the Evolutionary History and Global Expansion of *Candida auris* Using Population Genomic Analyses," *mBio* 11, no. 2 (2020): e03364-19.

34. Path Arora, Prerna Singh, Yue Wang, Anamika Yadav, Kalpana Pawar, Ashtosh Singh, Gadi Padmavati, et al., "Environmental Isolation of *Candida auris* from the Coastal Wetlands of Andaman Islands, India," *mBio* 12, no. 2 (2021): e03181-20.

35. Arturo Casadevall, Dimitrios P. Kontoyiannis, and Vincent Robert, "On the Emergence of *Candida auris*: Climate Change, Azoles, Swamps, and Birds," *mBio* 10, no. 4 (2019): e01397-19; Brendan R. Jackson, Nancy Chow, Kaitlin Forsberg, Anastasia P. Litvintseva, Shawn R. Lockhart, Rory Welsh, Snigdha Vallabhaneni, et al., "On the Origins of a Species: What Might Explain the Rise of *Candida auris*?," *Journal of Fungi* 5, no. 3 (2019): 58. The putative link between an increasing number of fungal infections and the warming climate has entered popular consciousness with the help of an HBO drama screened in 2023 called *The Last of Us*. The television series was referenced in an opinion article in the *New York Times*: Neil Vora, "'The Last of Us' Is Right: Our Warming Planet Is a Petri Dish," *New York Times*, April 6, 2023. For information

on mesophiles, see Sarah C. Watkinson, Lynne Boddy, and Nicholas P. Money, *The Fungi*, 3rd ed. (Amsterdam: Academic Press, 2016), 173–174. Changes in rainfall and other weather patterns rather than temperature may be more important in the spread of mycoses in some regions: Anil A. Panackal, "Global Climate Change and Infectious Diseases: Invasive Mycoses," *Journal of Earth Science and Climate Change* 1 (2011): 108.

36. Ewa Ksiezopolska and Toni Gabaldón, "Evolutionary Emergence of Drug Resistance in *Candida* Opportunistic Pathogens," *Genes* 9, no. 9 (2018): 461.

37. Lise N. Jørgensen and Thies M. Heick, "Azole Use in Agriculture, Horticulture, and Wood Preservation—Is It Indispensable?," *Frontiers in Cellular and Infection Microbiology* 11 (2021): 730297; Paul E. Verweij, Maiken C. Arendrup, Ana Alastruey-Izquierdo, Jeremy A. W. Gold, Shawn R. Lockhart, Tom Chiller, and P. Lewis White, "Dual Use of Antifungals in Medicine and Agriculture: How Do We Help Prevent Resistance Developing in Human Pathogens?," *Drug Resistance Updates* 65 (2022): 100885.

38. Ron Pinhasi, Boris Gasparian, Gregory Areshian, Diana Zardaryan, Alexia Smith, Guy Bar-Oz, and Thomas Higham, "First Direct Evidence of Chalcolithic Footwear from the Near Eastern Highlands," *PLoS ONE* 5, no. 6 (2010): e10984.

39. Contact lens solutions keep the eye clean with hydrogen peroxide, which works as a general disinfectant, and other compounds with more specific antimicrobial properties. The combination of natural cleansing and contact lens solutions works fine unless the cleaning fluids become contaminated with fungi. This is what happened in the United States in 2005 and 2006, when an outbreak of fungal keratitis affected 130 patients. One-third of the patients suffered eye damage that was serious enough to require corneal transplants. The CDC traced these cases to batches of contact lens solution manufactured by Bausch and Lomb, Inc., and legal settlements to victims have cost the company an estimated $1 billion. The fungus that caused this eye damage was a species of *Fusarium*, which normally grows on plants. Its spores must have landed in the lens solution during manufacture. Fungal keratitis continues to be a problem for wearers who are not careful to wash their lenses with fresh cleaning solutions. Y. Wang, H. Chen, T. Xia, and Y. Huang, "Characterization of Fungal Microbiota on Normal Ocular Surface of Humans," *Clinical Microbiology and Infection* 26, no. 1 (2020): 123.e9–123.e13; Sisinthy Shivaji, Rajagopalaboopathi Jayasudha, Gumpili S. Prashanthi, Kotakonda Arunasri, and Taraprasad Das, "Fungi of the Human Eye: Culture to Mycobiome," *Experimental Eye Research* 217 (2022): 108968; Arthur B. Epstein, "In the Aftermath of the *Fusarium* Keratitis Outbreak: What Have We Learned?," *Clinical Ophthalmology* 1, no. 4 (2007): 355–366.

40. A description of mycetoma of the foot in the three-thousand-year-old Indian *Atharvaveda* is the oldest record of a human mycosis; Ainsworth, *Introduction*, 1–2. Readers interested in this disease should consult Henry Vandyke Carter's book based on his observations in Bombay, where he served with the Indian Medical Service: *On Mycetoma; Or, the Fungus Disease of India* (London: J. & A. Churchill, 1874). Carter was the illustrator of *Gray's Anatomy*, and his hand-colored drawings of ferocious foot infections in *On Mycetoma* make this a collector's item.

41. Kristina Killgrove, Thomas Böni, and Francesco M. Galassi, "A Possible Case of Mycetoma in Ancient Rome (Italy, 2nd–3rd Centuries AD)," https://doi.org/10.31235/osf.io/2vjxk.

42. Bikash R. Behera, Sanjib Mishra, Manmath K. Dhir, Rabi N. Panda, and Sagarika Samantaray, "'Madura Head'—A Rare Case of Craniocerebral Maduromycosis," *Indian Journal of*

Neurosurgery 7 (2018): 159–163. Madura hand is another rare presentation of this mycosis: K. Rahman, M. Naim, and M. Farooqui, "Mycetoma of Hand—An Unusual Presentation," *Internet Journal of Dermatology* 8, no. 1 (2009), https://ispub.com/IJD/8/1/4863.

43. Rosane Orofino-Costa, Priscila M. de Macedo, Anderson M. Rodrigues, and Andréa R. Bernardes-Engemann, "Sporotrichosis: An Update on Epidemiology, Etiopathogenesis, Laboratory and Clinical Therapeutics," *Anais Brasileiros de Dermatologia* 92, no. 5 (2017): 606–620. Roses have prickles rather than thorns, so the infection mechanism for sporotrichosis involves a prickle prick rather than thorn prick, if we insist on the correct botanical definitions. Sporotrichosis is another example of a zoonotic mycosis that can be spread to humans from their pet cats.

44. Yvonne Gräser, Janine Fröhlich, Wolfgang Presber, and Sybren de Hoog, "Microsatellite Markers Reveal Geographic Population Differentiation in *Trichophyton rubrum*," *Journal of Medical Microbiology* 56, no. 8 (2007): 1058–1065; P. Zhan, K. Dukik, D. Li, J. Sun, J. B. Stielow, B. Gerrits van den Ende, B. Brankovics, et al., "Phylogeny of Dermatophytes with Genomic Character Evaluation of Clinically Distinct *Trichophyton rubrum* and *T. violaceum*," *Studies in Mycology* 89 (2018): 153–175.

45. "Athlete's Foot (Tinea Pedis) Treatment Market to Reach US$1.7 Bn by End of 2027," PharmiWeb.com, April 1, 2021, https://www.pharmiweb.com/press-release/2021-04-01/athlete-s-foot-tinea-pedis-treatment-market-to-reach-us-17-bn-by-end-of-2027.

CHAPTER THREE

1. My modest contributions to experimental mycology represent an extension of the pioneering studies on fungal spores by A. H. R. Buller (1874–1944) and Philip Gregory (1907–1986). Buller was the Einstein of mycology, and Gregory is known as the father of modern aerobiology, which is the study of spores and other airborne biological particles. Like me, Buller and Gregory suffered from asthma. By developing methods for measuring the concentrations of airborne spores, Gregory and his colleagues were responsible for drawing attention to fungi as a cause of allergy: Philip H. Gregory and John M. Hirst, "Possible Role of Basidiospores as Air-borne Allergens," *Nature* 170 (1952): 414. Asthma is not a qualification for spending decades studying spores. After all, one of the most influential mycologists of the twentieth century, C. T. Ingold (1905–2010), had no breathing issues, published papers on spores over a span of seventy years, and lived to the age of 104.

2. Alex Sakula, "Sir John Floyer's *A Treatise of the Asthma* (1698)," *Thorax* 39, no. 4 (1984): 248–254.

3. A spherical spore with a diameter 4 µm has a volume of 3.4×10^{-17} m³; 100,000 of these spores occupy a space of 3.4×10^{-12} m³. If these spores are dispersed evenly in one cubic meter of air, each spore will sail in a volume of air that is three hundred billion times larger than itself.

4. The inhalation of four hundred liters of air per hour (or 0.4 m³), with an average spore concentration of five thousand spores per cubic meter, over a lifespan of seventy-nine years, exposes the individual to 1.4 billion spores: $5,000 \text{ m}^{-3} \times 0.4 \text{ m}^3 \times 24 \times 365 \times 79 = 1.4 \times 10^9$ spores. The total volume of these spores, based on the volume of the individual spore (calculated from note 3 above), equals $1.4 \times 10^9 \times 3.4 \times 10^{-17} \text{ m}^3 = 4.8 \times 10^{-8} \text{ m}^3 = 4.8 \times 10^{-5} \text{ L} = 0.05$ mL. The density

of a spore is close to water, so the estimated mass of spores inhaled over a lifetime is 0.05 g or 50 milligrams, which is a bit lighter than a garden pea.

5. Paul Klenerman, an immunologist from the University of Oxford, provides a nice introduction to immunology: *The Immune System: A Very Short Introduction* (Oxford: Oxford University Press, 2018). The authoritative source on allergy is a weighty, two-volume book: A. Wesley Burks, Stephen T. Holgate, Robyn E. O'Hehir, David H. Broide, Leonard B. Bacharier, Gurjit K. Khurana Hershey, and R. Stokes Peebles, *Middleton's Allergy: Principles and Practice*, 9th ed. (Amsterdam: Elsevier, 2020).

6. Immunoglobulin E (IgE) is the antibody that plays a vital role in type I hypersensitivity reactions found in asthma and other allergic diseases. IgE is also a component of the immune reaction against parasitic worms. There is growing evidence that the innate immune system is also involved in asthma: Stephen T. Holgate, "Innate and Adaptive Immune Responses in Asthma," *Nature Medicine* 18 (2012): 673–683.

7. William E. Steavenson, *Spasmodic Asthma: A Thesis for the M.B. Degree of the University of Cambridge* (Cambridge: Deighton, Bell & Co., 1879).

8. Anon., "Obituary: William Edward Steavenson, M.D. Cantab., M.R.C.P.," *British Medical Journal* (June 6, 1891): 1261–1262. He died from influenza and bronchitis. Bronchitis is the most common complication of influenza. It is an inflammatory illness like asthma and shares the same type of antibody response involving immunoglobulin E (IgE): Christopher E. Brightling, "Chronic Cough Due to Nonasthmatic Eosinophilic Bronchitis: ACCP Evidence-Based Clinical Practice Guidelines," *Chest* 129, no. 1 suppl. (2006): 116S–121S.

9. Kathryn J. Waite, "Blackley and the Development of Hay Fever as a Disease of Civilization in the Nineteenth Century," *Medical History* 39, no. 2 (1995): 186–196.

10. David W. Denning, B. Ronan O'Driscoll, Cory M. Hogaboam, Paul Bowyer, and Robert M. Niven, "The Link between Fungi and Severe Asthma: A Summary of the Evidence," *European Respiratory Journal* 27, no. 2 (2006): 615–626; Gavin Dabrera, Virginia Murray, Jean Emberlin, Jonathan G. Ayres, Christopher Collier, Yoland Clewlow, and Patrick Sachon, "Thunderstorm Asthma: An Overview of the Evidence Base and Implications for Public Health Advice," *Quarterly Journal of Medicine* 106, no. 3 (2013): 207–217. The phenomenon of fungal-induced thunderstorm asthma was not recognized until 1983, when an asthma "epidemic" in Birmingham was linked to high levels of spores associated with a storm: G. E. Packe, P. S. Archer, and Jon G. Ayres, "Asthma and the Weather," *The Lancet* 322, no. 8344 (1983): 281; H. Morrow Brown and Felicity Jackson, "Asthma and the Weather," *The Lancet* 322, no. 8350 (1983): 630.

11. The grow and blow model was originally proposed for the dispersal of the fungus *Coccidioides*, but seems likely to apply to other species: James D. Tamerius and Andrew C. Comrie, "Coccidioidomycosis Incidence in Arizona Predicted by Seasonal Precipitation," *PLoS ONE* 6, no. 6 (2011): e21009.

12. Agnieszka Grinn-Gofroń and Agnieszka Strzelczak, "Changes in Concentration of *Alternaria* and *Cladosporium* Spores during Summer Storms," *International Journal of Biometeorology* 57, no. 5 (2013): 759–768; Ajay Kevat, "Thunderstorm Asthma: Looking Back and Looking Forward," *Journal of Asthma and Allergy* 13 (2020): 293–299; Nur S. Idrose, Shyamali C. Dharmage, Adrian J. Lowe, Katrina A. Lambert, Caroline J. Lodge, Michael J. Abramson, Jo A. Douglass,

et al., "A Systematic Review of the Role of Grass Pollen and Fungi in Thunderstorm Asthma," *Environmental Research* 181 (2020): 108911.

13. Mark Jackson, *Asthma: The Biography* (Oxford: Oxford University Press, 2009). Jackson is concerned with the social history of asthma rather than the science, but mention of fungal spores would not have been amiss.

14. Morell Mackensie, *Hay Fever and Paroxysmal Sneezing*, 4th ed. (London: J. & A. Churchill, 1887), 10.

15. Erich Wittkower and M. D. Berlin, "Studies in Hay-Fever Patients (the Allergic Personality)," *Journal of Mental Science* 84 (1938): 352–369. This paper was specific in its study of hay fever, as a seasonal allergy, but the wider concept of "the allergic personality" embraces the psychological characteristics of asthma patients.

16. Renee D. Goodwin, "Toward Improving Our Understanding of the Link between Mental Health, Lung Function, and Asthma Diagnosis. The Challenge of Asthma Measurement," *American Journal of Respiratory and Critical Care Medicine* 194, no. 11 (2016): 1313–1315.

17. Nicholas P. Money, *Carpet Monsters and Killer Spores: A Natural History of Toxic Mold* (New York: Oxford University Press, 2004).

18. Cornelia Witthauer, Andrew T. Gloster, Andrea H. Meyer, and Roselind Lieb, "Physical Diseases among Persons with Obsessive Compulsive Symptoms and Disorder: A General Population Study," *Social Psychiatry and Psychiatric Epidemiology* 49, no. 12 (2014): 2013–2022.

19. O. P. Sharma, "Marcel Proust (1871–1922): Reassessment of His Asthma and Other Maladies," *European Respiratory Journal* 15, no. 5 (2000): 958–960. Proust wrote much of his *In Search of Lost Time* in a cork-lined bedroom in Paris in an attempt to escape his invisible airborne enemies.

20. Paul Bowyer, Marcin Fraczek, and David W. Denning, "Comparative Genomics of Fungal Allergens and Epitopes Shows Widespread Distribution of Closely Related Allergen and Epitope Orthologues," *BMC Genomics* 7 (2006): 251; Viswanath P. Kurup and Banani Banerjee, "Fungal Allergens and Peptide Epitopes," *Peptides* 21, no. 4 (2000): 589–599.

21. Noah W. Palm, Rachel K. Rosenstein, and Ruslan Medzhitov, "Allergic Host Defences," *Nature* 484 (2012): 465–472; Michael Gross, "Why Did Evolution Give Us Allergies?," *Current Biology* 25, no. 2 (2015): R53–55; Alvaro Daschner and Juan González Fernández, "Allergy in an Evolutionary Framework," *Journal of Molecular Evolution* 88, no. 1 (2020): 66–76.

22. Grain silos packed with moldy barley create dense clouds of spores when they are unloaded, with one study showing a peak concentration of one billion spores per cubic meter of air: John Lacey, "The Microbiology of Moist Barley Storage in Unsealed Silos," *Annals of Applied Biology* 69, no. 3 (1971): 187–212. Sampling of the airborne dust during cereal harvesting in Lincolnshire in the 1970s showed a peak concentration of two hundred million spores per cubic meter of air: C. S. Darke, J. Knowelden, J. Lacey, and A. Milford Ward, "Respiratory Disease of Workers Harvesting Grain," *Thorax* 31, no. 2 (1976): 294–302. Twenty three percent of the farm workers in this study reported symptoms of wheezing and other respiratory complaints, but the remaining 77 percent of employees said that they were symptom-free. A similar spore count was recorded from a Swedish storehouse filled with wood chips used for fuel: Göran Blomquist, Gunnar Ström, and Lars-Helge Strömquist, "Sampling of High Concentrations of Airborne Fungi," *Scandinavian Journal of Work, Environment, and Health* 10, no. 2 (1984): 109–113. Other records of high spore concentrations include measurements of 128 million spores per

cubic meter of air in a Portuguese cork factory, forty million spores per cubic meter in Finnish cow barns, and twenty million spores per cubic meter in Norwegian sawmills: John Lacey, "The Air Spora of a Portuguese Cork Factory," *Annals of Occupational Hygiene* 16, no. 3 (1973): 223–230; Rauno Hanhela, Kyösti Louhelainen, and Anna-Liisa Pasanen, "Prevalence of Microfungi in Finnish Cow Barns and Some Aspects of the Occurrence of *Wallemia sebi* and *Fusaria*," *Scandinavian Journal of Work, Environment, and Health* 21, no. 3 (1994): 223–228; Wijnand Eduard, Per Sandven, and Finn Levy, "Exposure and IgG Antibodies to Mold Spores in Wood Trimmers: Exposure–Response Relationships with Respiratory Symptoms," *Applied Occupational and Environmental Hygiene* 9, no. 1 (1995): 44–48. A ten-year follow-up study of the workers exposed to the phenomenal levels of spores in the Norwegian sawmills found no evidence of long-term health effects: Karl Færden, May B. Lund, Trond M. Aaløkken, Wijnand Eduard, Per Søstrand, Sverre Langård, and Johny Kongerud, "Hypersensitivity Pneumonitis in a Cluster of Sawmill Workers: A 10-Year Follow-Up of Exposure, Symptoms, and Lung Function," *International Journal of Occupational and Environmental Health* 20, no. 2 (2014): 167–173. The use of dust masks has become routine since the original study of sawmills. The *Guinness World Records* refers to a global all-time record concentration of 194 million spores per cubic meter of air that was measured in Wales, but I have been unable to track down the source of this measurement: https://www.guinnessworldrecords.com/world-records/450409-largest-fungal-spore-count.

23. Lisa A. Reynolds and B. Brett Finlay, "Early Life Factors That Affect Allergy Development," *Nature Reviews Immunology* 17, no. 8 (2017): 518–528; B. Campbell, C. Raherison, C. J. Lodge, A. J. Lowe, T. Gislason, J. Heinrich, J. Sunyer, et al., "The Effects of Growing Up on a Farm on Adult Lung Function and Allergic Phenotypes: An International Population-Based Study," *Thorax* 72, no. 3 (2017): 236–244.

24. Andrew H. Liu, "Revisiting the Hygiene Hypothesis for Allergy and Asthma," *Journal of Allergy and Clinical Immunology* 136, no. 4 (2015): 860–865.

25. Money, *Carpet Monsters*.

26. Indoor molds seem to blossom in the vacuum created by the removal of the bacteria: Laura-Isobel McCall, Chris Callewaert, Qiyun Zhu, Se J. Song, Amina Bouslimani, Jeremiah J. Minich, Madeline Ernst, et al., "Home Chemical and Microbial Transitions across Urbanization," *Nature Microbiology* 5, no. 1 (2020): 108–115.

27. My parents tried this remedy for me without success. The dust mite allergen is described by Andy Chevigné and Alain Jacquet, "Emerging Roles of the Protease Allergen Der p 1 in House Dust Mite–Induced Airway Inflammation," *Journal of Allergy and Clinical Immunology* 142, no. 2 (2018): 398–400. Allergen avoidance as an asthma treatment is addressed by E. M. Rick, K. Woolnough, C. H. Pashley, and A. J. Wardlaw, "Allergic Fungal Airway Disease," *Journal of Investigational Allergology and Clinical Immunology* 26, no. 6 (2016): 344–354.

28. Keigo Kainuma, Akihiko Terada, Reiko Tokuda, Mizhuo Nagao, Nobuo Kubo, and Takao Fujisawa, "Wearing a Mask during Sleep Improved Asthma Control in Children," *Journal of Allergy and Clinical Immunology* 131 (2013): AB4; Barbara J. Polivka, Kamal Eldeirawi, Luz Huntington-Moskos, and Sharmilee M. Nyenhuis, "Mask Use Experiences, COVID-19, and Adults with Asthma: A Mixed-Methods Approach," *Journal of Allergy and Clinical Immunology: In Practice* 10, no. 1 (2022): 116–123. Face masks appear to be effective in reducing the symptoms of allergic rhinitis: Erdem Mengi, Cüneyt Orhan Kara, Uğur Alptürk, and Bülent Topuz, "The

Effect of Face Mask Usage on the Allergic Rhinitis Symptoms in Patients with Pollen Allergy during the Covid-19 Pandemic," *American Journal of Otolaryngology* 43, no. 1 (2022): 103206.

29. Eric K. Chu and Jeffrey M. Drazen, "Asthma: One Hundred Years of Treatment and Onward," *American Journal of Respiratory and Critical Care Medicine* 171, no. 11 (2005): 1202–1208.

30. Sheldon G. Cohen, "Asthma among the Famous: Roger E. C. Altounyan (1922–1987) British Physician and Pharmacologist," *Allergy and Asthma Proceedings* 19, no. 5 (1998): 328–332; Jack Howell, "Roger Altounyan and the Discovery of Cromolyn (Sodium Cromoglycate)," *Journal of Allergy and Clinical Immunology* 115, no. 4 (2005): 882–885.

31. Teresa To, Sanja Stanojevic, Ginette Moores, Andrea S. Gershon, Eric D. Bateman, Alvaro A. Cruz, and Louis-Phillipe Boulet, "Global Asthma Prevalence in Adults: Findings from the Cross-Sectional World Health Survey," *BMC Public Health* 12 (2012): 204; I. Asher and N. Pearce, "Global Burden of Asthma among Children," *International Journal of Tuberculosis and Lung Disease* 18, no. 11 (2014): 1269–1278.

32. Elizabeth H. Tham, Evelyn X. L. Loo, Yanan Zhu, and Lynette P.-C. Shek, "Effects of Migration on Allergic Diseases," *International Archives of Allergy and Immunology* 178 (2019): 128–140. Born in Birmingham, A. H. R. Buller (see note 1 above) escaped his asthma on the Canadian Prairies when he moved to Winnipeg in 1904 to found the Department of Botany at the University of Manitoba. He wrote that, "so far as the number of microorganisms is concerned, the climate of Central Canada during the winter must be one of the best in any civilised country in the world": Arthur H. R. Buller, and Charles W. Lowe, "Upon the Number of Micro-organisms in the Air of Winnipeg," *Transactions of the Royal Society of Canada*, ser. 3, 4 (1910): 41–58.

33. Daniel L. Hamilos, "Allergic Fungal Rhinitis and Rhinosinusitis," *Proceedings of the American Thoracic Society* 7, no. 3 (2010): 245–252; Peter Small, Paul K. Keith, and Harold Kim, "Allergic Rhinitis," *Allergy, Asthma, and Clinical Immunology* 14, suppl. 2 (2018): 51.

34. Ulrich Costabel, Yasunari Miyazaki, Annie Pardo, Dirk Koschel, Francesco Bonella, Paolo Spagnolo, Josune Guzman, et al., "Hypersensitivity Pneumonitis," *Nature Reviews Disease Primers* 6, no. 1 (2020): 65; J. Davidson, J. McErlane, K. Aljboor, S. L. Barratt, A. Jeyabalan, A. R. L. Medford, A. M. Borman, and H. Adamali, "Musical Instruments, Fungal Spores and Hypersensitivity Pneumonitis," *QJM* 112, no. 4 (2019): 287–289.

35. Bibek Paudel, Theodore Chu, Meng Chen, Vanitha Sampath, Mary Prunicki, and Kari C. Nadeau, "Increased Duration of Pollen and Mold Exposure Are Linked to Climate Change," *Scientific Reports* 11 (2021): 12816.

36. Michael R. Knowles and Richard C. Boucher, "Mucus Clearance as a Primary Innate Defense Mechanism for Mammalian Airways," *Journal of Clinical Investigation* 109, no. 5 (2002): 571–577; Ximena Bustamante-Marin and Lawrence E. Ostrowski, "Cilia and Mucociliary Clearance," *Cold Spring Harbor Perspectives in Biology* 9, no. 4 (2017): a028241.

37. Avani R. Patel, Amar R. Patel, Shivank Singh, Shantanu Singh, and Imran Khawaja, "Treating Allergic Bronchopulmonary Aspergillosis: A Review," *Cureus* 11, no. 4 (2019): e4538; Avani R. Patel, Amar R. Patel, Shivank Singh, Shantanu Singh, and Imran Khawaja, "Diagnosing Allergic Bronchopulmonary Aspergillosis: A Review," *Cureus* 11, no. 4 (2019): e4550.

38. Aaron S. Miller and Robert W. Wilmott, "The Pulmonary Mycoses," in *Kendig's Disorders of the Respiratory Tract in Children*, 9th ed., ed. Robert W. Wilmott, Andrew Bush, Robin R.

Deterding, Felix Ratjen, Peter Sly, Heather J. Zar, and Albert P. Li (Philadelphia: Elsevier, 2019), 507–527e3.

39. Pamela P. Lee and Yu-Lung Lau, "Cellular and Molecular Defects Underlying Invasive Fungal Infections—Revelations from Endemic Mycoses," *Frontiers in Immunology* 8 (2017): 735.

40. Russell E. Lewis and Dimitrios P. Kontoyiannis, "Invasive Aspergillosis in Glucocorticoid-Treated Patients," *Medical Mycology* 47, suppl. 1 (2009): S271–S281.

41. Tobias Lahmer, Silja Kriescher, Alexander Herner, Kathrin Rothe, Christoph D. Spinner, Jochen Schneider, Ulrich Mayer, et al., "Invasive Pulmonary Aspergillosis in Critically Ill Patients with Severe COVID-19 Pneumonia: Results from the Prospective AspCOVID-19 Study," *PLoS ONE* 16, no. 3 (2021): e0238825.

42. Shawn R. Lockhart, Mitsuru Toda, Kaitlin Benedict, Diego H. Caceres, and Anastasia P. Litvintseva, "Endemic and Other Dimorphic Mycoses in the Americas," *Journal of Fungi* 7 (2021): 151.

43. L. F. Shubitz, C. D. Butkiewicz, S. M. Dial, and C. P. Lindan, "Incidence of *Coccidioides* Infection among Dogs Residing in a Region in Which the Organism Is Endemic," *Journal of the American Veterinary Medical Association* 226, no. 11 (2005): 1846–1850.

44. The numbers are taken from Felix Bongomin, Sara Gago, Rita O. Oladele, and David W. Denning, "Global and Multi-National Prevalence of Fungal Diseases—Estimate Precision," *Journal of Fungi* 3 (2017): 57, which also serves as a useful source of data for other chapters.

45. "Fungal Diseases: Blastomycosis," CDC, accessed July 15, 2023, https://www.cdc.gov /fungal/diseases/blastomycosis/index.html; Katrina Thompson, Alana K. Sterkel, and Erin G. Brooks, "Blastomycosis in Wisconsin: Beyond the Outbreaks," *Academic Forensic Pathology* 7, no. 1 (2017): 119–129; Keith Matheny, "Fungal Infection Outbreak Affects 90+ Workers at Escanaba Paper Mill," *Detroit Free Press*, April 8, 2023.

46. P. Lewis White, Jessica S. Price, and Matthijs Backx, "*Pneumocystis jirovecii* Pneumonia: Epidemiology, Clinical Manifestation and Diagnosis," *Current Fungal Infection Reports* 13 (2019): 260–273; Gilles Nevez, Philippe M. Hauser, and Solène Le Gal, "*Pneumocystis jirovecii*," *Trends in Microbiology* 28, no. 12 (2020): 1034–1035; R. Benson Weyant, Dima Kabbani, Karen Doucette, Cecilia Lau, and Carlos Cervera, "*Pneumocystis jirovecii*: A Review with a Focus on Prevention and Treatment," *Expert Opinion on Pharmacotherapy* 22, no. 12 (2021): 1579–1592.

CHAPTER FOUR

1. Alon Tal, *Pollution in a Promised Land: An Environmental History of Israel* (Berkeley: University of California Press, 2002), 1–4.

2. Sandra C. Signore, Christoph P. Dohm, Gunter Schütze, Mathias Bähr, and Pawel Kermer, "*Scedosporium apiospermum* Brain Abscesses in a Patient after Near-Drowning—A Case Report with 10-Year Follow-Up and a Review of the Literature," *Medical Mycology Case Reports* 17 (2017): 17–19.

3. P. Hartmann, A. Ramseier, F. Gudat, M. J. Mihatsch, W. Polasek, and C. Geisenhoff, "Das Normgewicht des Gehirns beim Erwachsenen in Abhängigkeit von Alter, Geschlecht, Körpergröße und Gewicht," *Pathologe* 15 (1994): 165–170.

4. Karoll J. Cortez, Emmanuel Roilides, Flavio Quiroz-Telles, Joseph Meletiadis, Charalampos Antachopoulos, Tena Knudsen, Wendy Buchanan, et al., "Infections Caused by *Scedosporium* spp.," *Clinical Microbiology Reviews* 21, no. 1 (2008): 157–197.

5. P. A. Kowacs, C. E. Soares Silvado, S. Monteiro de Almeida, M. Ramos, K. Abrão, L. E. Madaloso, R. L. Pinheiro, et al., "Infection of the CNS by *Scedosporium apiospermum* after Near Drowning: Report of a Fatal Case and Analysis of Its Confounding Factors," *Journal of Clinical Pathology* 57 (2004): 205–207.

6. "Stop Neglecting Fungi," *Nature Microbiology* 2 (2017): 17120.

7. Felix Bongomin, Sara Gago, Rita O. Oladele, and David W. Denning, "Global and Multi-National Prevalence of Fungal Diseases—Estimate Precision," *Journal of Fungi* 3, no. 4 (2017): 57.

8. The fungus *Pneumocystis jirovecii* causes pneumocystis pneumonia in AIDS patients (see chapter 3).

9. "Fungal Disease: *C. neoformans* Infection Statistics," CDC, accessed July 15, 2023, https://www.cdc.gov/fungal/diseases/cryptococcosis-neoformans/statistics.html.

10. Priority for the claim that all fungi are opportunists seems to lie with Raymond Vanbreuseghem (1909–1993), who was a mycologist at the Institute for Tropical Medicine in Antwerp: R. Vanbreuseghem and C. de Vroey, "Systemic Opportunistic Fungal Infections," *Postgraduate Medical Journal* 55 (1979): 593–594.

11. Anuradha Chowdhary, Shallu Kathuria, Kshitij Agarwal, and Jacques F. Meis, "Recognizing Filamentous Basidiomycetes as Agents of Human Disease: A Review," *Medical Mycology* 52, no. 8 (2014): 782–797.

12. C. Correa-Martinez, A. Brentrup, K. Hess, K. Becker, A. H. Groll, and F. Schaumburg, "First Description of a Local *Coprinopsis cinerea* Skin and Soft Tissue Infection," *New Microbes and New Infections* 21 (2018): 102–104.

13. Erin L. Greer, Todd J. Kowalski, Monica L. Cole, Dylan V. Miller, and Larry M. Baddour, "Truffle's Revenge: A Pig-Eating Fungus," *Cardiovascular Pathology* 17, no. 5 (2008): 342–343.

14. Adela Enache-Angoulvant and Christophe Hennequin, "Invasive *Saccharomyces* Infection: A Comprehensive Review," *Clinical Infectious Diseases* 41, no. 11 (2005): 1559–1568. A strain of *Saccharomyces cerevisiae* that is used as a probiotic is implicated in many cases. Some yeast specialists regard this as a different species called *Saccharomyces boulardii*, although the distinction between species and strains is a matter of opinion rather than science in this instance. Rare cases of invasive disease caused by ordinary yeast strains used in baking have also been reported.

15. Arturo Casadevall and Liise-anne Pirofski, "The Damage-Response Framework of Microbial Pathogenesis," *Nature Reviews Microbiology* 1, no. 1 (2003): 17–24; Mary A. Jabra-Rizk, Eric F. Kong, Christina Tsui, M. Hong Nguyen, Cornelius J. Clancy, Paul L. Fidel, and Mairi Noverr, "*Candida albicans* Pathogenesis: Fitting within the Host-Microbe Damage Response Framework," *Infection and Immunity* 84, no. 10 (2016): 2724–2739; Antonis Rokas, "Evolution of the Human Pathogenic Lifestyle in Fungi," *Nature Microbiology* 7, no. 5 (2022): 607–619.

16. Arturo Casadevall, "Determinants of Virulence in the Pathogenic Fungi," *Fungal Biology Reviews* 21, no. 4 (2007): 130–132; Cene Gostinčar, Janja Zajc, Metka Lenassi, Ana Plemenitaš, Sybren de Hoog, Abdullah M. S. Al-Hatmi, and Nina Gunde-Cimerman, "Fungi between Extremotolerance and Opportunistic Pathogenicity on Humans," *Fungal Diversity* 93 (2018): 195–213.

17. One of the fungi blackened with melanin that causes impromptu brain infections has another complicated Latin name: this is *Cladophialophora bantiana*. To begin pronouncing Latin names of species you should speak the syllables out loud, clay-doe-fi-al- and so on, slowly at first, then repeat the chain faster and you will soon sound as seductive as a Roman bard. *Cladophialophora* is a soil fungus with a global distribution that forms lovely velvety colonies when it is grown in a culture dish. This fungus is especially worrying because it infects people with intact immune systems and kills about 70 percent of its victims. Early symptoms of infection include headaches, seizure, arm pain, and ataxia—or loss of muscle coordination. The fungus produces chains of spores that become airborne, and so we assume that it gets into us through the lungs or nasal passages. We have no idea why this ubiquitous fungus infects a tiny fraction of the people who must come in contact with its spores all the time. Studies on the immune systems of patients suggest that they may have an underlying vulnerability that would not be noticed if they had not been diagnosed with this fungus, but that is all we know. Treatments are limited to the surgical removal of infected tissue and use of powerful antifungal drugs, but the high mortality figures speak for themselves. This is a very unpleasant fungus: Todd P. Levin, Darric E. Baty, Thomas Fekete, Allan L. Truant, and Byungse Suh, "*Cladophialophora bantiana* Brain Abscess in a Solid-Organ Transplant Recipient: Case Report and Review of the Literature," *Journal of Clinical Microbiology* 42, no. 9 (2004): 4374–4378; Jon Velasco and Sanjay Revankar, "CNS Infections Caused by Brown-Black Fungi," *Journal of Fungi* 5, no. 3 (2019): 60.

18. Patient-to-patient transmission of pneumocystis pneumonia seems to be an exception among the mycoses (see chapter 3).

19. Emily Monosson, *Blight: Fungi and the Coming Pandemic* (New York: W. W. Norton, 2023).

20. Synnecrosis means dying together: José P. Veiga, "Commensalism, Amensalism, and Synnecrosis," in *The Encyclopedia of Evolutionary Biology*, vol. 1, ed. Richard M. Kliman (Oxford: Academic Press, 2016), 322–328. All biology is a battle, *bellum omnium contra omnes*, as Hobbes said. There is no generosity in nature, only nightmares in the making, as I say on rare occasions when the charms of the fungi fail to sweeten my view of life.

21. Peter G. Pappas, "Cryptococcal Infections in Non-HIV-Infected Patients," *Transactions of the American Clinical and Climatological Association* 124 (2013): 61–79.

22. Judith N. Steenbergen, Howard Shuman, and Arturo Casadevall, "*Cryptococcus neoformans* Interactions with Amoebae Suggest an Explanation for Its Virulence and Intracellular Pathogenic Strategy in Macrophages," *Proceedings of the National Academy of Sciences USA* 98, no. 26 (2001): 15245–15250; Rhys A. Watkins, Alexandre Andrews, Charlotte Wynn, Caroline Barisch, Jason S. King, and Simon A. Johnston, "*Cryptococcus neoformans* Escape from *Dictyostelium* Amoeba by Both WASH-Mediated Constitutive Exocytosis and Vomocytosis," *Frontiers in Cellular and Infection Microbiology* 8 (2018): 108.

23. Liliana Scorzoni, Ana C. A. de Paula e Silva, Caroline M. Marcos, Patricia A. Assato, Wanessa C. M. A. de Melo, Haroldo C. de Oliveira, Caroline B. Costa-Orlandi, et al., "Antifungal Therapy: New Advances in the Understanding and Treatment of Mycosis," *Frontiers in Microbiology* 8 (2017): 36.

24. "Fungal Disease: *C. neoformans* Infection," CDC, accessed July 15, 2023, https://www.cdc .gov/fungal/diseases/cryptococcosis-neoformans/index.html; World Health Organization,

WHO Fungal Priority Pathogens List to Guide Research, Development and Public Health Action (Geneva: World Health Organization, 2022); Abbygail C. Spencer, Katelyn R. Brubaker, and Sylvie Garneau-Tsodikova, "Systemic Fungal Infections: A Pharmacist/Researcher Perspective," *Fungal Biology Reviews* 44 (2023): 100293. A relative of *Cryptococcus neoformans* called *Cryptococcus gattii* also causes serious brain infections and is proficient at doing so in people with perfectly healthy immune systems: Lamin Saidykhan, Chinaemerem U. Onyishi, and Robert C. May, "The *Cryptococcus gattii* Species Complex: Unique Pathogenic Yeasts with Understudied Virulence Mechanisms," *PLoS Neglected Tropical Diseases* 16, no. 12 (2022): e0010916.

25. Dimitrios P. Kontoyiannis, Hongbo Yang, Jinlin Song, Sneha S. Kelkar, Xi Yang, Nkechi Azie, Rachel Harrington, et al., "Prevalence, Clinical and Economic Burden of Mucormycosis-Related Hospitalizations in the United States: A Retrospective Study," *BMC Infectious Diseases* 16 (2016): 730.

26. Sylvia Slaughter, "Love Endures in the Face of Sorrow," *The Tennessean*, January 12, 2003, pp. 6–13.

27. Mahnoor Sukaina, "Re-Emergence of Mucormycosis in COVID-19 Recovered Patients Transiting from Silent Threat to an Epidemic in India," *JoGHR* 5 (2021): e2021067; Neil Stone, Nitin Gypta, and Ilan Schwartz, "Mucormycosis: Time to Address This Deadly Fungal Infection," *Lancet Microbe* 2, no. 8 (2021): e343–e344.

28. Jana M. Ritter, Atis Muehlenbachs, Dianna M. Blau, Christopher D. Paddock, Wun-Ju Shieh, Clifton P. Drew, Brigid C. Batten, et al., "*Exserohilum* Infections Associated with Contaminated Steroid Injections: A Clinicopathologic Review of 40 Cases," *American Journal of Pathology* 183, no. 3 (2013): 881–892.

29. Diana Pisa, Ruth Alonso, Alberto Rábano, Izaskun Rodal, and Luis Carrasco, "Different Brain Regions Are Infected with Fungi in Alzheimer's Disease," *Scientific Reports* 5 (2015): 15015; Ruth Alonso, Diana Pisa, Ana M. Fernández-Fernández, and Luis Carrasco, "Infection of Fungi and Bacteria in Brain Tissue from Elderly Persons and Patients with Alzheimer's Disease," *Frontiers in Aging Neuroscience* 10 (2018): 159.

30. Bodo Parady, "Innate Immune and Fungal Model of Alzheimer's Disease," *Journal of Alzheimer's Disease Reports* 2, no. 1 (2018): 139–152; Yifan Wu, S. Du, J. L. Johnson, H.-Y. Tung, C. T. Landers, Y. Liu, B. G. Seman, et al., "Microglia and Amyloid Precursor Protein Coordinate Control of Transient *Candida* Cerebritis with Memory Deficits," *Nature Communications* 10 (2019): 58.

31. Kelly Servick, doi:10.1126/science.aaw0147; R. C. Roberts, C. B. Farmer, and C. K. Walker, "The Human Brain Microbiome: There Are Bacteria in Our Brains!," paper presented at the Neuroscience 2018 Conference, November 6, https://www.abstractsonline.com/pp8/#!/4649/presentation/32057.

32. Ruth Alonso, Diana Pisa, Ana Fernández-Fernández, Alberto Rábano, and Luis Carrasco, "Fungal Infection in Neural Tissue of Patients with Amyotrophic Lateral Sclerosis," *Neurobiology of Disease* 108 (2018): 249–260.

33. Diana Pisa, Ruth Alonso, and Luis Carrasco, "Parkinson's Disease: A Comprehensive Analysis of Fungi and Bacteria in Brain Tissue," *International Journal of Biology Sciences* 16, no. 7 (2020): 1135–1152.

34. Mary Duenwald, "Parkinson's 'Clusters' Getting a Closer Look," *New York Times*, May 14, 2002.

CHAPTER FIVE

1. Yang Sun, Tao Zuo, Chun P. Cheung, Wenxi Gu, Yating Wan, Fen Zhang, Nan Chen, et al., "Population-Level Configurations of Gut Mycobiome across 6 Ethnicities in Urban and Rural China," *Gastroenterology* 160, no. 1 (2021): 272–286.

2. The *Candida* species in the Chinese study was *Candida dubliniensis*. This gut fungus, discovered in Ireland in the 1990s, has a global distribution.

3. Emily A. Speakman, Ivy M. Dambuza, Fabián Salazar, and Gordon D. Brown, "T Cell Antifungal Immunity and the Role of C-Type Lectin Receptors," *Trends in Immunology* 41, no. 1 (2020): 61–76.

4. Lu Wu, Tiansheng Zeng, Massimo Deligios, Luciano Milanesi, Morgan G. I. Langille, Angelo Zinellu, Salvatore Rubino, et al., "Age-Related Variation of Bacterial and Fungal Communities in Different Body Habitats across the Young, Elderly, and Centenarians in Sardinia," *mSphere* 5, no. 1 (2020): e00558-19.

5. Andrea K. Nash, Thomas A. Auchtung, Matthew C. Wong, Daniel P. Smith, Jonathan R. Gesell, Matthew C. Ross, Christopher J. Stewart, et al., "The Gut Mycobiome of the Human Microbiome Project Healthy Cohort," *Microbiome* 5, no. 1 (2017): 153.

6. Mubanga H. Kabwe, Surendra Vikram, Khodani Mulaudzi, Janet K. Jansson, and Thulani P. Makhalanyane, "The Gut Mycobiota of Rural and Urban Individuals Is Shaped by Geography," *BMC Microbiology* 20, no. 1 (2020): 257.

7. Eric van Tilburg Bernardes, Veronika K. Pettersen, Mackensie W. Gutierrez, Isabelle Laforest-Lapointe, Nicholas G. Jendzjowsky, Jean-Baptiste Cavin, Fernando A. Vicentini, et al., "Intestinal Fungi Are Causally Implicated in Microbiome Assembly and Immune Development in Mice," *Nature Communications* 11, no. 1 (2020): 2577; Tahliyah S. Mims, Qusai A. Abdallah, Justin D. Stewart, Sydney P. Watts, Catrina T. White, Thomas V. Rousselle, Ankush Gosain, et al., "The Gut Mycobiome of Healthy Mice Is Shaped by the Environment and Correlates with Metabolic Outcomes in Response to Diet," *Communications Biology* 4, no. 1 (2021): 281.

8. Katherine D. Mueller, Hao Zhang, Christian R. Serrano, R. Blake Billmyre, Eun Y. Huh, Philipp Wiemann, Nancy P. Keller, et al., "Gastrointestinal Microbiota Alteration Induced by *Mucor circinelloides* in a Murine Model," *Journal of Microbiology* 57, no. 6 (2019): 509–520.

9. M. Mar Rodríguez, Daniel Pérez, Felipe J. Chaves, Eduardo Esteve, Pablo Marin-Garcia, Gemma Xifra, Joan Vendrell, et al., "Obesity Changes the Human Gut Mycobiome," *Scientific Reports* 5 (2015): 14600.

10. William D. Fiers, Iris H. Gao, and Iliyan D. Iliev, "Gut Mycobiota Under Scrutiny: Fungal Symbionts or Environmental Transients?," *Current Opinion in Microbiology* 50 (2019): 79–86.

11. Mario Matijašić, Tomislav Meštrović, Hana Čipčić Paljetak, Mihaela Perić, Anja Barešić, and Donatella Verbanac, "Gut Microbiota beyond Bacteria—Mycobiome, Virome, Archaeome, and Eukaryotic Parasites in IBD," *International Journal of Molecular Sciences* 21 (2020): 2668; Umang Jain, Aaron M. Ver Heul, Shanshan Xiong, Martin H. Gregory, Elora G. Demers, Justin T. Kern, Chin-Wen Lai, et al., "*Debaryomyces* Is Enriched in Crohn's Disease Intestinal Tissue and Impairs Healing in Mice," *Science* 371 (2021): 1154–1159. There has been a lot of interest in a putative link between antibodies called ASCAs produced in response to proteins in the cell wall

of baker's yeast, *Saccharomyces cerevisiae*, and the development of Crohn's disease: Heba N. Is-kandar and Matthew A. Ciorba, "Biomarkers in Inflammatory Bowel Disease: Current Practices and Recent Advances," *Translational Research* 159, no. 4 (2012): 313–325. Some research shows that although these antibodies play a role in gut inflammation they are not correlated with the consumption of the dietary yeast in baked foods and beer: Anne S. Kvehaugen, Martin Aas-brenn, and Per G. Farup, "Anti-*Saccharomyces cerevisiae* Antibodies (ASCA) Are Associated with Body Fat Mass and Systemic Inflammation, But Not with Dietary Yeast Consumption: A Cross-Sectional Study," *BMC Obesity* 4 (2017): 28.

12. Irina Leonardi, Sudarshan Paramsothy, Itai Doron, Alexa Semon, Nadeem O. Kaakoush, Jose C. Clemente, Jeremiah J. Faith, et al., "Fungal Trans-Kingdom Dynamics Linked to Re-sponsiveness to Fecal Microbiota Transplantation (FMT) Therapy in Ulcerative Colitis," *Cell Host and Microbe* 27, no. 5 (2020): 823–829.

13. Arthur C. Macedo, André O. V. de Faria, and Pietro Ghezzi, "Boosting the Immune System, from Science to Myth: Analysis [of] the Infosphere with Google," *Frontiers in Medicine* 6 (2019): 165.

14. Yao Zuo, Hui Zhan, Fen Zhang, Qin Liu, Eugene Y. K. Tso, Grace C. Y. Lui, Nan Chen, et al., "Alterations in Fecal Fungal Microbiome of Patients with COVID-19 during Time of Hospitalization until Discharge," *Gastroenterology* 159, no. 4 (2020): 1302–1310.

15. Bing Zhai, Mihaela Ola, Thierry Rolling, Nicholas L. Tosini, Sar Joshowitz, Eric R. Littmann, Luigi A. Amoretti, et al., "High-Resolution Mycobiota Analysis Reveals Dynamic Intestinal Translocation Preceding Invasive Candidiasis," *Nature Medicine* 26 (2020): 59–64; Bastian Seelbinder, Jiarui Chen, Sasha Brunke, Ruben Vazquez-Uribe, Rakesh Santhaman, Anne-Christin Meyer, Felipe Senne de Oliveira Lino, et al., "Antibiotics Create a Shift from Mutualism to Competition in Human Gut Communities with a Longer-Lasting Impact on Fungi Than Bacteria," *Microbiome* 8 (2020): 133.

16. Sara Botschuijver, Guus Roeselers, Evgeni Levin, Daisy M. Jonkers, Olaf Welting, Sig-rid E. M. Heinsbroek, Heleen H. de Weerd, et al., "Intestinal Fungal Dysbiosis Is Associated with Visceral Hypersensitivity in Patients with Irritable Bowel Syndrome and Rats," *Gastroen-terology* 153, no. 4 (2017): 1026–1039.

17. Natalia Vallianou, Dimitris Kounatidis, Gerasimos Socrates Christodoulatos [such a great name that I had to waive the citation style of middle name initial only], Fotis Panagopoulos, Irene Karampela, and Maria Dalamaga, "Mycobiome and Cancer: What Is the Evidence?," *Can-cers* 13 (2021): 3149.

18. Berk Aykut, Smruit Pushalkar, Ruonan Chen, Qianhao Li, Raquel Abengozar, Jacque-line I. Kim, Sorin A. Shadaloey, et al., "The Fungal Mycobiome Promotes Pancreatic Oncogen-esis via Activation of MBL," *Nature* 574 (2019): 264–267; Jessica R. Galloway-Peña and Dimi-trios P. Kontoyiannis, "The Gut Mycobiome: The Overlooked Constituent of Clinical Outcomes and Treatment Complications in Patients with Cancer and Other Immunosuppres-sive Conditions," *PLoS Pathogens* 16, no. 4 (2020): e1008353; Lian Narunsky-Haziza, Gregory D. Sepich-Poore, Ilana Livyatan, Omer Asraf, Cameron Martino, Deborah Nejman, Nancy Gavert, et al., "Pan-Cancer Analyses Reveal Cancer-Type-Specific Fungal Ecologies and Bacteriome Interactions," *Cell* 185, no. 20 (2022): 3789–3806; Anders B. Dohlman, Jared Klug, Marissa Mesko, Iris H. Gao, Steven M. Lipkin, Xiling Shen, and Iliyan D. Iliev, "A Pan-Cancer

Mycobiome Analysis Reveals Fungal Involvement in Gastrointestinal and Lung Tumors," *Cell* 185, no. 20 (2022): 3807–3822.

19. Nicholas P. Money, "Hyphal and Mycelial Consciousness: The Concept of the Fungal Mind," *Fungal Biology* 125 (2021): 257–259.

20. Ecologists use the term *ecotype* to describe a population of a species of plant or animal that is adapted to a local environment. An ecotype is a variant within a species. *Mycotype* is used in a different way to describe a community of fungi that is identified by the presence of single species of fungus. *Enterotype* is another term used to distinguish between different versions of the gut microbiome based on their bacterial composition.

21. B. P. Krom, S. Kidwai, and J. M. Ten Cate, "Candida and Other Fungal Species: Forgotten Players of Healthy Oral Microbiota," *Journal of Dental Research* 93, no. 5 (2014): 445–451; B. Y. Hong, A. Hoare, A. Cardenas, A. K. Dupuy, L. Choquette, A. L. Salner, P. K. Schauer, et al., "The Salivary Mycobiome Contains 2 Ecologically Distinct Mycotypes," *Journal of Dental Research* 99, no. 6 (2020): 730–738.

22. M. N. Zakaria, M. Furuta, T. Takeshita, Y. Shibata, R. Sundari, N. Eshima, T. Ninomiya, et al., "Oral Mycobiome in Community-Dwelling Elderly and Its Relation to Oral and General Health Conditions," *Oral Diseases* 23, no. 7 (2017): 973–982; Eefje A. Kraneveld, Mark J. Buijs, Marc J. Bonder, Marjolein Visser, Bart J. F. Keijser, Wim Crielaard, and Egija Zaura, "The Relation between Oral *Candida* Load and Bacterial Microbiome Profiles in Dutch Older Adults," *PLoS ONE* 7, no. 8 (2012): e42770. Changes in the oral mycobiome associated with dentures were superimposed on a huge disparity between the baseline levels of *Candida* in the two populations: the saliva of the Japanese patients contained an average of ten thousand *Candida* cells per milliliter, compared with up to one hundred million yeast cells in the same volume of spit from the Dutch patients. It is possible that this discrepancy was due to the use of different DNA primers in these studies.

23. David W. Denning, Matthew Kneale, Jack D. Sobel, and Riina Rautemaa-Richardson, "Global Burden of Recurrent Vulvovaginal Candidiasis: A Systematic Review," *Lancet Infectious Diseases* 18, no. 11 (2018): e339–e347; Brett A. Tortelli, Warren G. Lewis, Jennifer E. Allsworth, Nadum Member-Meneh, Lynne R. Foster, Hilary E. Reno, Jeffrey F. Peipert, et al., "Associations between the Vaginal Microbiome and *Candida* Colonization in Women of Reproductive Age," *American Journal of Obstetrics and Gynecology* 222, no. 5 (2020): 471.e1–e9.

24. Ning-Ning Liu, Xingping Zhao, Jing-Cong Tan, Sheng Liu, Bo-Wen Li, Wang-Xing Xu, Lin Peng, et al., "Mycobiome Dysbiosis in Women with Intrauterine Adhesions," *Microbiology Spectrum* 10, no. 4 (2022): e0132422.

25. Erik van Tilburg Bernardes, Mackenzie W. Gutierrez, and Marie-Claire Arrieta, "The Fungal Microbiome and Asthma," *Frontiers in Cellular and Infection Microbiology* 10 (2020): 583418.

26. Raphaël Enaud, Renaud Prevel, Eleonora Ciarlo, Fabien Beaufils, Gregoire Wieërs, Benoit Guery, and Laurence Delhaes, "The Gut-Lung Axis in Health and Respiratory Diseases: A Place for Inter-Organ and Inter-Kingdom Crosstalks," *Frontiers in Cellular and Infection Microbiology* 10 (2020): 9.

27. Tomasz Gosiewski, Dominika Salamon, Magdalena Szopa, Agnieska Sroka, Maciej T. Malecki, and Malgorzata Bulanda, "Quantitative Evaluation of Fungi of the Genus *Candida* in

the Feces of Adult Patients with Type 1 and 2 Diabetes—A Pilot Study," *Gut Pathogens* 6 (2014): 43; A. M. Yang, T. Inamine, K. Hochrath, P. Chen, L. Wang, C. Llorente, S. Bluemel, et al., "Intestinal Fungi Contribute to Development of Alcoholic Liver Disease," *Journal of Clinical Investigations* 127, no. 7 (2017): 2829–2841; Lu Jiang, Peter Stärkel, Jian-Gao Fan, Derrick E. Fouts, Petra Bacher, and Bernd Schnabl, "The Gut Mycobiome: A Novel Player in Chronic Liver Diseases," *Journal of Gastroenterology* 56, no. 1 (2021): 1–11.

28. Jessica D. Forbes, Charles N. Bernstein, Helen Tremlett, Gary Van Domselaar, and Natlaie C. Knox, "A Fungal World: Could the Gut Mycobiome Be Involved in Neurological Disease?," *Frontiers in Microbiology* 9 (2019): 3249; Saumya Shah, Albertu Locca, Yair Dorsett, Claudia Cantoni, Laura Ghezzi, Qingqi Lin, Suresh Bokoliya, et al., "Alterations of the Gut Mycobiome in Patients with MS," *EBioMedicine* 71, no. 1 (2021): 103557.

29. Mahmoud Ghannoum with Eve Adamson, *Total Gut Balance: Fix Your Mycobiome Fast for Complete Digestive Wellness* (Woodstock, VT: Countryman Press, 2019).

30. M. Ghannoum, C. Smith, E. Adamson, N. Isham, I. Salem, and M. Retuerto, "Effect of Mycobiome Diet on Gut Fungal and Bacterial Communities of Healthy Adults," *Journal of Probiotics and Health* 8, no. 1 (2020): 215.

31. Kearney T. W. Gunsalus, Stephanie N. Tornberg-Belanger, Nirupa R. Matthan, Alice H. Lichtenstein, and Carol A. Kumamoto, "Manipulation of Host Diet to Reduce Gastrointestinal Colonization by the Opportunistic Pathogen *Candida albicans*," *mSphere* 1, no. 1 (2015): e00020-15.

CHAPTER SIX

1. The genus *Penicillium* was named by Heinrich Friedrich Link in 1809, who described the spore stalks or conidiophores produced from the mycelium as *fertilibus erectis apice penicillatis*, meaning raised fertile [branches] with brush-like tips: Heinrich F. Link, "Observationes in Ordines Plantarum Naturales: Dissertatio Ima," *Gesellschaft Naturforschender Freunde zu Berlin Magazin* 3, no. 1 (1809): 3–42.

2. *Pencillium* evolved in the Cretaceous. This is the timing that we infer from the DNA clocks in multiple species of *Penicillium* that appear to have been ticking for more than seventy million years: Jacob L. Steenwyk, Xing-Xing Shen, Abigail L. Lind, Gustavo H. Goldman, and Antonis Rokas, "A Robust Phylogenomic Time Tree for Biotechnologically and Medically Important Fungi in the Genera *Aspergillus* and *Penicillium*," *mBio* 10 (2019): e00925-19.

3. Frank Maixner, Mohamed S. Sarhan, Kun D. Huang, Adrian Tett, Alexander Schoenafinger, Stefania Zingale, Aitor Blanco-Míguez, et al., "Hallstatt Miners Consumed Blue Cheese and Beer During the Iron Age and Retained a Non-Westernized Gut Microbiome until the Baroque Period," *Current Biology* 31, no. 23 (2021): 5149–5162.

4. Nathaniel J. Dominy, "Ferment in the Family Tree," *Proceedings of the National Academy of Sciences USA* 112, no. 2 (2015): 308–309; Nicholas P. Money, *The Rise of Yeast: How the Sugar Fungus Shaped Civilization* (Oxford: Oxford University Press, 2018).

5. Jiajing Wang, Leping Jiang, and Hanlong Sun, "Early Evidence for Beer Drinking in a 9000-Year-Old Platform Mound in Southern China," *PLoS ONE* 16, no. 8 (2021): e0255833. Jiajing Wang and colleagues identified microfossils of filamentous fungi and yeast in the pottery

remains. Filamentous fungi are used as starters in rice wine fermentation to break down starch into sugars, and yeast feeds on the sugars, producing alcohol. Incidentally, rice wine is really rice beer because it is made from grains that contain starch that is converted into sugars in the first step of the fermentation, called saccharification. Wines are made from grape must and other fruit juices, which are full of sugars so that yeast can get to work without this saccharification step.

6. Laure Segurel, Perle Guarino-Vignon, Nina Marchi, Sophie Lafosse, Romain Laurent, Céline Bon, Alexandre Fabre, et al., "Why and When Was Lactase Persistence Selected For? Insights from Central Asian Herders and Ancient DNA," *PLoS Biology* 18, no. 6 (2020): e3000742; William T. T. Taylor, Julia Clark, Jamranjav Bayarsaikhan, Tumurbaatar Tuvshinjargal, Jessica T. Jobe, William Fitzhugh, Richard Kortum, et al., "Early Pastoral Economies and Herding Transitions in Eastern Eurasia," *Scientific Reports* 10 (2020): 1001; Mélanie Salque, Peter I. Bogucki, Joanna Pyzel, Iwona Sobkowiak-Tabaka, Ryszard Grygiel, Marzena Szmyt, and Richard P. Evershed, "Earliest Evidence for Cheese Making in the Sixth Millennium BC in Northern Europe," *Nature* 493 (2013): 522–525.

7. Pliny, *Natural History*, trans. Harris Rackham, Loeb Classical Library 353 (Cambridge, MA: Harvard University Press, 1942), Book XI, XCVII, 582–585, lines 240–242; Petronius, *Satyricon*, trans. Michael Heseltine, rev. Eric H. Warmington, Loeb Classical Library 15 (Cambridge, MA: Harvard University Press, 1987), 148–149, line 66. The cheese description in the *Satyricon* comes from Habinnas, a guest at the feast of Trimalchio, who is asked about an earlier dinner.

8. Emilie Dumas, Alice Feurtey, Ricardo C. Rodríguez de la Vega, Stéphanie Le Prieur, Alodie Snirc, Monika Coton, Anne Thierry, et al., "Independent Domestication Events in the Blue-Cheese Fungus *Penicillium roqueforti*," *Molecular Ecology* 29 (2020): 2639–2660.

9. Jeanne Ropars, Estelle Didiot, Ricardo C. Rodríguez de la Vega, Bastien Bennetot, Monika Coton, Elisabeth Poirier, Emmanuel Coton, et al., "Domestication of the Emblematic White Cheese-Making Fungus *Penicillium camemberti* and Its Diversification into Two Varieties," *Current Biology* 30, no. 22 (2020): 4441–4453, e1–e4.

10. Marie-Christine Montel, Solange Buchin, Adrien Mallet, Céline Delbes-Paus, Dominique A. Vuitton, Nathalie Desmasures, and François Berthier, "Traditional Cheeses: Rich and Diverse Microbiota with Associated Benefits," *International Journal of Food Microbiology* 177 (2014): 136–154.

11. Eric Dugat-Bony, Lucille Garnier, Jeremie Denonfoux, Stéphanie Ferreira, Anne-Sophie Sarthou, Pascal Bonnarme, and Françoise Irlinger, "Highlighting the Microbial Diversity of 12 French Cheese Varieties," *International Journal of Food Microbiology* 238 (2016): 265–273.

12. Yuanchen Zhang, Erik K. Kastman, Jeffrey S. Guasto, and Benjamin E. Wolfe, "Fungal Networks Shape Dynamics of Bacterial Dispersal and Community Assembly in Cheese Rind Microbiomes," *Nature Communications* 9 (2018): 336.

13. Clifton Fadiman, *Any Number Can Play* (Cleveland, OH: World Publishing, 1957), 105. In the same book (106), Fadiman described Roquefort as "Ewe-born, cave-educated, [and] perfected by moldy bread."

14. Montel et al., "Traditional Cheeses." Raw milk is enriched in vitamins that are lost in pasteurization, contains a healthier mixture of fats than processed milk (according to some

nutritionists), and may even confer some protection against the development of asthma and other allergies in children.

15. Thibault Caron, Mélanie Le Piver, Anne-Claire Péron, Pascale Lieben, René Lavigne, Sammy Brunel, Daniel Roueyre, et al., "Strong Effect of *Penicillium roqueforti* Populations on Volatile and Metabolic Compounds Responsible for Aromas, Flavor and Texture in Blue Cheeses," *International Journal of Food Microbiology* 354 (2021): 109174.

16. B. G. J. Knols and R. De Jong, "Limburger Cheese as an Attractant for the Malaria Mosquito *Anopheles gambiae* s.s.," *Parasitology Today* 12, no. 54 (1996): 159–161.

17. Monika Coton, Franck Deniel, Jérôme Mounier, Rozenn Joubrel, Emeline Robieu, Audrey Pawtowski, Sabine Jeuge, et al., "Microbial Ecology of French Dry Fermented Sausages and Mycotoxin Risk Evaluation during Storage," *Frontiers in Microbiology* 12 (2021): 737140. Concerns have been raised about the possibility of mycotoxin contamination of cheeses, but there have been no proven cases of poisoning associated with cheese consumption: Alan D. W. Dobson, "Mycotoxins in Cheese," in *Cheese: Chemistry, Physics and Microbiology*, 4th ed., ed. Paul L. H. McSweeney, Patrick F. Fox, Paul D. Cotter, and David W. Everett (London: Academic Press, 2017), 595–601.

18. Giancarlo Perrone, Robert A. Samson, Jens C. Frisvad, Antonia Susca, Nina Gunde-Cimerman, Filomena Epifani, and Jos Houbraken, "*Penicillium salamii*, A New Species Occurring during Seasoning of Dry-Cured Meat," *International Journal of Food Microbiology* 193 (2015): 91–98.

19. Andrea Osimani, Ilario Ferrocino, Monica Agnolucci, Luca Cocolin, Manuela Giovannetti, Caterina Cristani, Michela Palla, et al., "Unveiling *Hákarl*: A Study of the Microbiota of the Traditional Icelandic Fermented Fish," *Food Microbiology* 82 (2019): 560–572. Most of the sharks are killed as bycatch, and their great age adds to this tragedy: Greenland sharks are the longest-lived vertebrates, with a maximum estimated life span approaching four hundred years.

20. There are frequent comparisons between the smell of *surströmming* and open sewers on the internet. This delicacy is one of the exhibits that can be tasted at the Disgusting Food Museum in Malmö (https://disgustingfoodmuseum.com/). Fermented fish dishes from Asia are described in the following review article: Yutika Narzary, Sandeep Nas, Arvind K. Goyal, Su S. Lam, Hermen Sarma, and Dolikajyoti Sharma, "Fermented Fish Products in South and Southeast Asian Cuisine: Indigenous Technology Processes, Nutrient Composition, and Cultural Significance," *Journal of Ethnic Foods* 8 (2021): 33.

21. David Downie, "A Roman Anchovy's Tale," *Gastronomica* 3 (2003): 25–28; Brian Keogh, *The Secret Sauce: A History of Lea & Perrins* (Worcester, UK: Leaper Books, 1997).

22. Kotaro Ito and Asahi Matsuyama, "Koji Molds for Japanese Soy Sauce Brewing: Characteristics and Key Enzymes," *Journal of Fungi* 7 (2021): 658.

23. M. J. Robert Nout and Kofi E. Aidoo, "Asian Fungal Fermented Food," in *The Mycota*, vol. 10, *Industrial Applications*, ed. Martin Hofrichter (Berlin: Springer, 2010), 29–58.

24. Climate may help to explain why the *Mucor* infections of humans described in chapter 4 are more common in India and other parts of Asia than Europe.

25. Money, *The Rise of Yeast*, 52.

26. Jack A. Whittaker, Robert I. Johnson, Tim J. A. Finnigan, Simon V. Avery, and Paul S. Dyer, "The Biotechnology of Quorn Mycoprotein: Past, Present and Future Challenges," in

Grand Challenges in Fungal Biotechnology, ed. Helena Nevalainen (Cham, Switzerland: Springer International Publishing, 2020), 59–79.

27. Pedro F. Souza Filho, Dan Andersson, Jorge A. Ferreira, and Mohammad J. Taherzadeh, "Mycoprotein: Environmental Impact and Health Aspects," *World Journal of Microbiology and Biotechnology* 35, no. 10 (2019): 147; Maurizio Cellura, Maria A. Cusenza, Sonia Longo, Le Q. Luu, and Thomas Skurk, "Life Cycle Environmental Impacts and Health Effects of Protein-Rich Food as Meat Alternatives: A Review," *Sustainability* 14 (2022): 979; Florian Humpenöder, Benjamin L. Bodirsky, Isabelle Weindl, Hermann Lotze-Campen, Tomas Linder, and Alexander Popp, "Projected Environmental Benefits of Replacing Beef with Microbial Protein," *Nature* 605, no. 7908 (2022): 90–96.

28. Robert King, Neil A. Brown, Martin Urban, and Kim E. Hammond-Kosack, "Inter-Genome Comparison of the Quorn Fungus *Fusarium venenatum* and the Closely Related Plant Infecting Pathogen *Fusarium graminearum*," *BMC Genomics* 19 (2018): 269.

29. The market for fungal products is dominated by yeast. See Nicholas P. Money, "The Fungus That's Worth $900 Billion a Year," *OUPblog*, February 25, 2018, https://blog.oup.com /2018/02/fungus-worth-900-billion/.

30. The energy value of gilled mushrooms varies from 22 to 31 calories per 100 grams for raw white button mushrooms to 44 calories per 100 grams of shiitake; 100 grams of romaine lettuce contains 20 calories. Measurements of the calorific value of truffles vary between studies and for different truffle species, but the high energy value of these fungi relative to gilled mushrooms is consistent. A study from China, for example, measured 378 calories per 100 grams of three species of *Tuber* from Yunnan, which matches the calorific value of Roquefort cheese. See U.S. Department of Agriculture, "Mushrooms, White, Raw," April 1, 2019, https://fdc.nal.usda.gov /fdc-app.html#/food-details/169251/nutrients; Xiangyuan Yan, Yanwei Wang, Xiaoyu Sang, and Li Fan, "Nutritional Value, Chemical Composition and Antioxidant Activity of Three *Tuber* Species from China," *AMB Express* 7, no. 1 (2017): 136.

CHAPTER SEVEN

1. U. Peintner, R. Pöder, and T. Pümpel, "The Iceman's Fungi," *Mycological Research* 102, no. 10 (1998): 1153–1162.

2. Luigi Capasso, "5300 Years Ago, the Ice Man Used Natural Laxatives and Antibiotics," *The Lancet* 352, no. 9143 (1998): 1864. Capasso's work was refuted by Håkan Tunón and Ingvar Svanberg, "Laxatives and the Ice Man," *The Lancet* 353, no. 9156 (1999): 925–926, who wrote, "Ethnobotanical data from preindustrial Northern Europe show that the fungus has had several non-medical uses, such as to protect metal blades from rust, to sharpen razors, as toys, floats or pincushions. So it is odd that Capasso concludes that the fungi kept by the Ice Man were used to treat a worm infection and not for any other purpose. . . . We find it astonishing that Capasso draws so many conclusions from such a limited amount of data."

3. Powerful drugs, including ivermectin, which was made famous during the COVID-19 pandemic, paralyze and kill the worms, and modern sanitation allows us to avoid the worms in the first place. Insouciant attitudes toward intestinal parasites are among the unearned privileges of today's affluence that must be judged naive against the global burden of billions of active

infections by hookworms, roundworms, and Ötzi's whipworm: Rachel L. Pullan, Jennifer L. Smith, Rashmi Jasrasaria, and Simon J. Brooker, "Global Numbers of Infection and Disease Burden of Soil Transmitted Helminth Infections in 2010," *Parasites Vectors* 7 (2014): 37.

4. Ulrike Grienke, Margit Zöll, Ursula Peintner, and Judith M. Rollinger, "European Medicinal Polypores—A Modern View on Traditional Uses," *Journal of Ethnopharmacology* 154, no. 3 (2014): 564–583.

5. Robert A. Blanchette, "*Haploporus odorus*: A Sacred Fungus in Traditional Native American Culture of the Northern Plains," *Mycologia* 89, no. 2 (1997): 233–240.

6. Investors view the medicinal mushroom industry as fragmented, meaning that hundreds of companies share the market in different countries. Decentralization can be good for consumers and provides plenty of opportunities for small-scale entrepreneurs to develop new product lines. This contrasts with the market for prescription and over-the-counter drugs, which is controlled by a few very powerful pharmaceutical companies. See "Global Mushroom Market (2020 to 2025)—Global Industry Trends, Share, Size, Growth, Opportunity and Forecast—Research-AndMarkets.com," Business Wire, July 1, 2020, https://www.businesswire.com/news/home/20200701005442/en/Global-Mushroom-Market-2020-to-2025--Global-Industry-Trends-Share-Size-Growth-Opportunity-and-Forecast--ResearchAndMarkets.com; Allana Akhtar, "5 'Functional' Mushrooms the Wellness Industry Is Obsessed with, from Lion's Mane to Turkey Tail," YahooMoney, April 7, 2022, https://money.yahoo.com/5-functional-mushrooms-wellness-industry-135455865.html.

7. *Cordyceps* is an ascomycete, more closely related to yeast than gilled mushrooms, and chaga is a mass of fungal tissues that does not produce any spores.

8. "Health Benefits of Mushrooms," WebMD, September 12, 2022, https://www.webmd.com/diet/health-benefits-mushrooms; "What Is the Nutritional Value of Mushroom Powder?," *Om* (blog), May 11, 2021, https://ommushrooms.com/blogs/blog/nutritional-value-of-mushroom-powder-m2.

9. Koichiro Mori, Yutaro Obara, Mitsuru Hirota, Yoshihito Azumi, Satomi Kinugasa, Satoshi Inatomi, and Norimichi Nakahata, "Nerve Growth Factor-Inducing Activity of *Hericium erinaceus* in 1321N1 Human Astrocytoma Cells," *Biological and Pharmaceutical Bulletin* 31, no. 9 (2008): 1727–1732; Mari Shimbo, Hirokazu Kawagishi, and Hidehiko Yokogoshi, "Erinacine A Increases Catecholamine and Nerve Growth Factor Content in the Central Nervous System of Rats," *Nutrition Research* 25, no. 6 (2005): 617–623. Although these are brief reports, they are the best publications on the effects of lion's mane on cultured nerve cells and rat brains. Most of the published studies on *Hericium* would never pass peer review in reliable scientific journals. One detailed analysis of the fungus looked promising: Hsing-Chun Kuo, Chien-Chien Lu, Chien-Heng Shen, Shui-Yi Tung, Meng Chiao Hsieh, Ko-Chao Lee, Li-Ya Li, et al., "*Hericium erinaceus* Mycelium and Its Isolated Erinacine A Protection from MPTP-Induced Neurotoxicity through the ER Stress, Triggering an Apoptosis Cascade," *Journal of Translational Medicine* 19 (2021): 67. I used the past tense, looked, because the study was retracted when the editors of the journal learned that the research was associated with a Taiwanese company called Grape King Bio, Ltd., which produces extracts from the mushroom.

10. Koichiro Mori, Satoshi Inatomi, Kenzi Ouchi, Yoshihito Azumi, and Takasi Tuchida, "Improving Effects of the Mushroom Yamabushitake (*Hericium erinaceus*) on Mild Cognitive

Impairment: A Double-Blind Placebo-Controlled Clinical Trial," *Phytotherapy Research* 23, no. 3 (2009): 367–372.

11. Tero Isokauppila, *Healing Mushrooms: A Practical and Culinary Guide to Using Mushrooms for Whole Body Health* (New York: Avery, 2017).

12. "Lion's Mane Capsules," FungiPerfecti, accessed July 15, 2023, https://fungi.com/products /lions-mane-capsules.

13. "Top 5 Lions Mane Health Benefits for Managing Erectile Dysfunction Effectively," Cure My Erectile Dysfunction, accessed July 15, 2023, https://curemyerectiledysfunction.com/top-5 -lions-mane-health-benefits-for-managing-erectile-dysfunction-effectively; "Lion's Mane Can Reduce Your Libido/Sex-Drive," *Boost Your Biology* (blog), August 17, 2020, https://www .ergogenic.health/blog/lions-mane-can-decrease-your-libido-sex-drive.

14. Hidde P. van Steenwijk, Aalt Bast, and Alie de Boer, "Immunomodulating Effects of Fungal Beta-Glucans: From Traditional Use to Medicine," *Nutrients* 13 (2021): 1333.

15. Kurt Buchmann, "Evolution of Innate Immunity: Clues from Invertebrates via Fish to Mammals," *Frontiers in Immunology* 5 (2014): 459.

16. Kenji Ina, Takae Kataoka, and Takafumi Ando, "The Use of Lentinan for Treating Gastric Cancer," *Anti-cancer Agents in Medicinal Chemistry* 13, no. 5 (2013): 681–688.

17. Yiran Zhang, Meng Zhang, Yifei Jiang, Xiulian Li, Yanli He, Pengjiao Zeng, Zhihua Guo, et al., "Lentinan as an Immunotherapeutic for Treating Lung Cancer: A Review of 12 Years Clinical Studies in China," *Journal of Cancer Research and Clinical Oncology* 144 (2018): 2177–2186.

18. "Medical Health Benefits of Beta-Glucans in Medicinal Mushrooms," WENY News, July 20, 2021, https://www.weny.com/story/44338597/medical-health-benefits-of-beta -glucans-in-medicinal-mushrooms; Christopher Hertzog, *Beta Glucan: A 21st Century Miracle?* (Bangkok: Booksmango, 2014).

19. Djibril M. Ba, Xiang Gao, Joshua Muscat, Laila Al-Shaar, Vernon Chinchilli, Xinyuan Zhang, Paddy Ssentongo, et al., "Association of Mushroom Consumption with All-Cause and Cause-Specific Mortality among American Adults: Prospective Cohort Study Findings from NHANES III," *Nutrition Journal* 20, no. 1 (2021): 38.

20. Djibril M. Ba, Xiang Gao, Laila Al-Shaar, Joshua E. Muscat, Vernon M. Chinchilli, Robert B. Beelman, and John P. Richie, "Mushroom Intake and Depression: A Population-Based Study Using Data from the US National Health and Nutrition Examination Survey (NHANES), 2005–2016," *Journal of Affective Disorders* 294 (2021): 686–692; Djibril M. Ba, Paddy Ssentongo, Robert B. Beelman, Joshua Muscat, Xiang Gao, and John P. Richie, "Higher Mushroom Consumption Is Associated with Lower Risk of Cancer: A Systematic Review and Meta-Analysis of Observational Studies," *Advances in Nutrition* 12, no. 5 (2021): 1691–1704.

21. Piotr Rzymski, "Comment on 'Mushroom Intake and Depression: A Population-Based Study Using Data from the US National Health and Nutrition Examination Survey (NHANES), 2005–2016,'" *Journal of Affective Disorders* 295 (2021): 937–938.

22. Chayakrit Krittanawong, Ameesh Isath, Joshua Hahn, Zhen Wang, Sonya E. Fogg, Dhrubajyoti Bandyopadhyay, Hani Jneid, et al., "Mushroom Consumption and Cardiovascular Health: A Systematic Review," *American Journal of Medicine* 134, no. 5 (2021): 637–642.e2.

23. Nicholas P. Money, "Are Mushrooms Medicinal?," *Fungal Biology* 120, no. 4 (2016): 449–453.

24. Christopher Hitchens, *God Is Not Great: How Religion Poisons Everything* (New York: Twelve, 2009), 150.

25. In addition to the web pages referring to the curative powers of the mushroom, many of the "shiitake acne" and "shiitake asthma" sites describe severe skin allergies in some people who consume the raw mushroom and in workers in the mushroom industry who handle the fruit bodies during packaging.

26. John Gerard, *The Herball, or, Generall Historie of Plantes*, 2nd ed., enlarged and amended by Thomas Johnson (London: Adam Islip, Joice Norton, and R. Whitakers, 1633), 1578, 1583; Horace, *Satires, Epistles, and Ars Poetica*, trans. H. Rushton Fairclough, Loeb Classical Library 194 (Cambridge, MA: Harvard University Press, 1929), *Satires* Book II, IV, 188–189, lines 20–21.

27. The study of medicinal mushrooms is lost in a madhouse of misrepresentation and pseudoscience that includes crackpot cures for terminal illnesses. For light relief, I nominate Robert Rogers, registered herbalist and author of *Mushroom Essences: Vibrational Healing from the Kingdom Fungi* (Berkeley, CA: North Atlantic Books, 2016), for The Batshit Crazy Award in Mycology. Rogers claims that mushrooms "express energy fields," which can be channeled by skilled practitioners to "help peel away the steel bars of long-held emotional and mental imprisonment." There are many contenders for the award, but a sentence from the blurb of the book by Roger should satisfy the judges: "Similar to flower essences, but made under a lunar cycle, mushroom essences work subtly to bring deep healing to the mind and body; they are particularly well suited for working with the 'shadow' or unintegrated parts of the psyche." Ötzi would have slapped Robert with his birch conks.

28. Won C. Bak, Ji H. Park, Yong A. Park, and Kang H. Ka, "Determination of Glucan Contents in the Fruiting Bodies and Mycelia of *Lentinula edodes* Cultivars," *Mycobiology* 42, no. 3 (2014): 301–304; Juan Chen, Xu Zeng, Yan L. Yang, Yong M. Xing, Qi Zhang, Jia Li, Ke Ma, et al., "Genomic and Transcriptomic Analyses Reveal Differential Regulation of Diverse Terpenoid and Polyketides Secondary Metabolites in *Hericium erinaceus*," *Scientific Reports* 7, no. 1 (2017): 10151; Marcus Künzler. "How Fungi Defend Themselves against Microbial Competitors and Animal Predators," *PLoS Pathogens* 14, no. 9 (2018): e1007184. Some medicinal mushroom companies choose to highlight these distinctions and emphasize that they are selling extracts from fruit bodies rather than mycelia. Others suggest that mycelia are superior sources of medicinals to mushrooms, and still more ignore the potential difference in chemistry between the two sources. In the end, neither claim affects consumers because the active compounds are never specified. Uncertainties about the marketing of extracts from fruit bodies versus mycelia is one aspect of wider concerns about the labeling of fungal products as foods and alternative medicines. A DNA barcoding study of different food products containing "wild mushrooms" revealed that many contained common cultivated mushrooms, and that some of the ingredient labels misrepresented the species of fungi in dried powders, soups, and pasta sauces: W. Dalley Cutler II, Alexander J. Bradshaw, and Bryn T. M. Dentinger, "What's for Dinner This Time? DNA Authentication of 'Wild Mushrooms' in Food Products Sold in the USA," *PeerJ* 2, no. 9 (2021): e11747.

29. Kenneth D. Clevenger, Jin W. Bok, Rosa Ye, Galen P. Miley, Maria H. Verdan, Thomas Velk, Cynthia Chen, et al., "A Scalable Platform to Identify Fungal Secondary Metabolites and

Their Gene Clusters," *Nature Chemical Biology* 13, no. 8 (2017): 895–901; Claudio Greco, Nancy P. Keller, and Antonis Rokas, "Unearthing Fungal Chemodiversity and Prospects for Drug Discovery," *Current Opinion in Microbiology* 51 (2019): 22–29; Matthew T. Robey, Lindsay K. Caesar, Milton T. Drott, Nancy P. Keller, and Neil L. Kelleher, "An Interpreted Atlas of Biosynthetic Gene Clusters from 1,000 Fungal Genomes," *Proceedings of the National Academy of Sciences USA* 118, no. 19 (2021): e2020230118; Kirstin Scherlach and Christian Hertweck, "Mining and Unearthing Hidden Biosynthetic Potential," *Nature Communications* 12 (2021): 3864.

30. Carsten Gründemann, Jakob K. Reinhardt, and Ulricke Lindequist, "European Medicinal Mushrooms: Do They Have Potential for Modern Medicine?—An Update," *Phytomedicine* 66 (2020): 153131.

31. Ravinder Kumar and Piyush Kumar, "Yeast-Based Vaccines: New Perspective in Vaccine Development and Application," *FEMS Yeast Research* 19, no. 2 (2019): foz007.

32. I read a book about bird's nest fungi (the only one on this subject) as a student and was struck with the intricate design of these things. Later, when I found them for the first time in Colorado, I felt something of "the tide of emotion" experienced by Stendhal in the Basilica di Santa Croce in Florence, where he visited the tombs of Machiavelli and Galileo, and saw the chiaroscuro frescos of Volterrano: "As I emerged from the porch of Santa Croce . . . I walked in constant fear of falling to the ground." The French author's response has been memorialized in a psychosomatic condition called Stendhal's syndrome that describes tourists swooning before great works of art. This diagnosis should be extended to people with an exceptional sensitivity toward the fungi: "*Sanctus stercore*," I thought in English when I gazed upon the tiny nests of the species whose Latin name is *Cyathus stercoreus*. Pursuing my mycological expression of Stendhal's syndrome, I anticipate feeling quite emotional if I am fortunate enough to visit the Basilica Santa Croce, where, beneath the fresco, lies the tomb of Pier Antonio Micheli (1679–1737). Micheli is celebrated as the father of experimental mycology for his experiments with mushroom spores, described in his magnum opus, *Nova Plantarum Genera*, published in 1729. There is a striking statue of Micheli in the colonnade outside the Uffizi, and he is also memorialized in street names in Florence and Rome. Sources: Harold J. Brodie, *The Bird's Nest Fungi* (Toronto: University of Toronto Press, 1975); Stendhal, *Rome, Naples and Florence*, trans. Richard N. Coe (Richmond, UK: John Calder, 1959), 301–302; Iain Bamforth, "Stendhal's Syndrome," *British Journal of General Practice* 60, no. 581 (2010): 945–946.

33. Olchowecki's original observations on the antibiotic stimulated the doctoral research of another student, Bhavdish Narain Johri, whose dissertation was the foundation for all the subsequent work on the cyathins: B. N. Johri, H. J. Brodie, A. D. Allbutt, W. A. Ayer, and H. Taube, "A Previously Unknown Antibiotic Complex from the Fungus *Cyathus helenae*," *Experientia* 27 (1971): 853; A. D. Allbutt, W. A. Ayer, H. J. Brodie, B. N. Johri, and H. Taube, "Cyathin, a New Antibiotic Complex Produced by *Cyathus helenae*," *Canadian Journal of Microbiology* 17, no. 11 (1971): 1401–1407. Harold Brodie wrote the book on the bird's nest fungi that I read as a student titled *The Bird's Nest Fungi* (Toronto: University of Toronto Press, 1975), and introduced a crumb of mirth, intentionally or otherwise, into an otherwise dry scientific article on mycelial mergers with the subheading, "Attempts at Mating with *Cyathus olla*." Less than a crumb.

34. Emma Dixon, Tatiana Schweibenz, Alison Hight, Brian Kang, Allyson Dailey, Sarah Kim, Meng-Yang Chen, et al., "Bacteria-Induced Static Batch Fungal Fermentation of the Diterpenoid

Cyathin A3, a Small-Molecule Inducer of Nerve Growth Factor," *Journal of Industrial Microbiology and Biotechnology* 38, no. 5 (2011): 607–615; Christian Bailly and Jin-Ming Gao, "Erinacine A and Related Cyathane Diterpenoids: Molecular Diversity and Mechanisms Underlying Their Neuroprotection and Anticancer Activities," *Pharmaceutical Research* 159 (2020): 104953.

CHAPTER EIGHT

1. "Celebratory Meal a Near Death Experience," *Raglan Chronicle*, May 9, 2020, https://www.raglanchronicle.co.nz/the-chronicle/2020/05/celebratory-meal-a-near-death-experience/; John Weekes, "Waikato Doctor Nearly Dies after Death Cap Mushroom Poisoning," *Stuff*, May 11, 2020, https://www.stuff.co.nz/national/health/121464993/waikato-doctor-nearly-dies-after-death-cap-mushroom-poisoning.

2. William E. Brandenburg and Karlee J. Ward, "Mushroom Poisoning Epidemiology in the United States," *Mycologia* 110, no. 4 (2018): 637–641; Jeremy A. W. Gold, Emily Kiernan, Michael Yeh, Brendan R. Jackson, and Kaitlin Benedict, "Health Care Utilization and Outcomes Associated with Accidental Poisonous Mushroom Ingestions—United States, 2016–2018," *MMWR Morbidity and Mortality Weekly Report* 70 (2021): 337–341. Between 1999 and 2016, more than seven thousand Americans were poisoned by mushrooms every year, with 60 percent of cases reported for children younger than six, and few resulting in more than brief gastrointestinal distress. During this period, there were seven or fewer fatalities per year from mushroom poisoning, which was comparable to the number of lethal snake bites.

3. Anne Pringle and Else C. Vellinga, "Last Chance to Know? Using Literature to Explore the Biogeography and Invasion Biology of the Death Cap Mushroom *Amanita phalloides* (Vaill. ex Fr.:Fr.) Link," *Biological Invasions* 8 (2006): 1131–1144; Anne Pringle, Rachel I. Adams, Hugh B. Cross, and Thomas D. Bruns, "The Ectomycorrhizal Fungus *Amanita phalloides* Was Introduced and Is Expanding Its Range on the West Coast of North America," *Molecular Ecology* 18 (2009): 817–833.

4. Here are the corresponding Latin names: oyster mushrooms, *Pleurotus ostreatus*; lion's mane, *Hericium erinaceus*; common puffballs, *Lycoperdon perlatum*; giant puffballs, *Calvatia gigantea*; golden chanterelles, *Cantharellus cibarius*; and porcini, ceps, or king boletes, *Boletus edulis*.

5. Dennis R. Benjamin, *Mushrooms: Poisons and Panaceas—A Handbook for Naturalists, Mycologists, and Physicians* (New York: W. H. Freeman & Co., 1995). *Amanita vaginata* is the grisette; *Amanita rubescens* is the blusher; the fool's mushroom is *Amanita verna*; and the destroying angels are *Amanita bisporigera*, *Amanita ocreata*, and *Amanita virosa*.

6. Britt A. Barnyard, "The Real Story behind Increased Amanita Poisonings in North America," *FUNGI Magazine* 8, no. 3 (2015): 6–9.

7. Chad Hyatt, *The Mushroom Hunter's Kitchen: Reimaging Comfort Food with a Chef Forager* (San Jose, CA: Chestnut Fed Books, 2018), 107–109.

8. Nicholas P. Money, *Mushrooms: A Natural and Cultural History* (London: Reaktion Books, 2017), 137–138.

9. I refer readers interested in mushroom conservation to a prescient essay whose publication attracted a great deal of baseless dissent by mushroomers: Nicholas P. Money, "Why Picking Wild Mushrooms May Be Bad Behaviour," *Mycological Research* 109, no. 2 (2005): 131–135.

10. Paolo Scocco, Giampietro Rupolo, and Diego De Leo, "Failed Suicide by *Amanita phalloides* (Mycetismus) and Subsequent Liver Transplant: Case Report," *Archives of Suicide Research* 4 (1998): 201–206.

11. Ismail Yilmaz, Fatih Ermis, Ilgaz Akata, and Ertugrul Kaya, "A Case Study: What Doses of *Amanita phalloides* and Amatoxins Are Lethal to Humans?," *Wilderness and Environmental Medicine* 26, no. 4 (2015): 491.

12. Yongzhuang Ye and Zhenning Liu, "Management of *Amanita phalloides* Poisoning: A Literature Review and Update," *Journal of Critical Care* 46 (2018): 17–22; Juliana Garcia, Vera M. Costa, Alexandra Carvalho, Paula Baptista, Paula G. de Pinho, Maria de Lourdes Bastos, and Félix Carvalho, "*Amanita phalloides* Poisoning: Mechanisms of Toxicity and Treatment," *Food and Chemical Toxicology* 86 (2015): 41–55. Death caps contain three groups of toxins: amatoxins, phallotoxins, and vomitoxins.

13. The lethal dose of alpha-amanitin is estimated to be 0.1–0.3 milligrams per kilogram body weight (from Yilmaz et al., "A Case Study," 491–496), which compares with 300–500 milligrams per kilogram for aspirin. Incidentally, alpha-amanitin is ten thousand times less deadly than botulinum toxin, or Botox, with an LD50 of 30 nanograms per kilogram. LD50 is the amount of a substance that kills half of the laboratory animals in an experiment. These estimates refer to oral administration.

14. Patrick L. West, Janet Lindgren, and B. Zane Horowitz, "*Amanita smithiana* Mushroom Ingestion: A Case of Delayed Renal Failure and Literature Review," *Journal of Medical Toxicology* 5, no. 1 (2009): 32–38.

15. Brandon Landry, Jeannette Whitton, Anna L. Bazzicalupo, Oldriska Ceska, and Mary L. Berbee, "Phylogenetic Analysis of the Distribution of Deadly Amatoxins among the Little Brown Mushrooms of the Genus *Galerina*," *PLoS ONE* 16, no. 2 (2021): e0246575.

16. Julian White, Scott A. Weinstein, Luc De Haro, Regis Bédry, Andreas Schaper, Bary H. Rumack, and Thomas Zilker, "Mushroom Poisoning: A Proposed New Clinical Classification," *Toxicon* 157 (2019): 53–65.

17. Regis Bedry, Isabelle Baudrimont, Gerard Deffieux, Edmond E. Creppy, Jean P. Pomies, Jean M. Ragnaud, Michel Dupon, et al., "Wild-Mushroom Intoxication as a Cause of Rhabdomyolysis," *New England Journal of Medicine* 345 (2001): 798–802.

18. Piotr Rzymski and Piotr Klimaszyk, "The Yellow Knight Fights Back: Toxicological, Epidemiological, and Survey Studies Defend Edibility of *Tricholoma equestre*," *Toxins* 10, no. 11 (2018): 468.

19. A one-kilogram potato contains between 20 and 130 milligrams of solanine hydrochloride, and mice studies provide an LD50 estimate of 42 milligrams of solanine per kilogram body weight. Using these numbers, a human would need to eat more than 20 kilograms of potatoes to absorb a comparable dose. Poisonings have occurred among people who have eaten more modest quantities of potatoes containing exceptionally high levels of solanine, which can develop when the tubers become spoiled in storage. Potatoes belong to a toxic family of plants that includes deadly nightshade, which carries a lethal dose of atropine in a few of its onyx-black berries. See National Center for Biotechnology Information, "PubChem Compound Summary for CID 118796405, Solanine HCl," accessed July 17, 2023, https://pubchem.ncbi.nlm.nih.gov/compound/Solanine-HCl.

20. Petteri Nieminen and Anne-Mari Mustonen, "Toxic Potential of Traditionally Consumed Mushroom Species—A Controversial Continuum with Many Unanswered Questions," *Toxins* 12, no. 10 (2020): 639.

21. Even morels upset some people: Benjamin, *Mushrooms*, 278.

22. Hikoto Ohta, Daisuke Watanabe, Chie Nomura, Daichi Saito, Koichi Inoue, Hajime Miyaguchi, Shuichi Harada, et al., "Toxicological Analysis of Satratoxins, the Main Toxins in the Mushroom *Trichoderma cornu-damae*, in Human Serum and Mushroom Samples by Liquid Chromatography–Tandem Mass Spectrometry," *Forensic Toxicology* 39 (2021): 101–113.

23. Fungi with coral shapes belong to the basidiomycetes and the ascomycetes. The fire coral is an ascomycete, whereas the hundreds of species of *Clavaria* or fairy clubs, *Ramaria*, and other "clavarioid" fungi, are basidiomycetes.

24. Luis E. Alonso-Aguilar, Adriana Montoya, Alejandro Kong, Arturo Estrada-Torres, and Roberto Garibay-Orijel, "The Cultural Significance of Wild Mushrooms in San Mateo Huexoyucan, Tlaxcala, Mexico," *Journal of Ethnobiology and Ethnomedicine* 10 (2014): 27.

25. "*Ramaria flava* (Schaeff.) Quél.," First Nature, accessed July 15, 2023, https://www.first-nature.com/fungi/ramaria-flava.php; Pamela M. North, *Poisonous Plants and Fungi in Colour* (London: Blandford Press, 1967), 109–110.

26. Charles McIlvaine, *One Thousand American Fungi: How to Select and Cook the Edible; How to Distinguish and Avoid the Poisonous* (Indianapolis, IN: Bowen-Merrill Co., 1900). I have celebrated the extraordinary life and career of Captain McIlvaine in an earlier book: *Mushrooms: A Natural and Cultural History* (London: Reaktion Books, 2017), 84–86.

27. Normal Mier, Sandrine Canete, Alain Klaebe, Luis Chavant, and Didier Fournier, "Insecticidal Properties of Mushroom and Toadstool Carpophores," *Phytochemistry* 41, no. 5 (1996): 1293–1299.

28. Paul A. Horgen, Allan C. Vaisius, and Joseph F. Ammirati, "The Insensitivity of Mushroom Nuclear RNA Polymerase Activity to Inhibition by Amatoxins," *Archives of Microbiology* 118 (1978): 317–319.

29. Frank M. Dugan, *Fungi in the Ancient World: How Mushrooms, Mildews, Molds, and Yeast Shaped the Early Civilizations of Europe, the Mediterranean, and the Near East* (St. Paul, MN: APS Press, 2008).

30. The literature on ergotism is voluminous. The following pair of papers on the Norwegian history of the phenomenon are relevant to outbreaks of ergotism in other regions: Torbjørn Alm and Brita Elvevåg, "Ergotism in Norway, Part 1: The Symptoms and Their Interpretation from the Late Iron Age to the Seventeenth Century," *History of Psychiatry* 24, no. 1 (2013): 15–33, and "Ergotism in Norway, Part 2: The Symptoms and Their Interpretation from the Eighteenth Century Onwards," *History of Psychiatry* 24, no. 2 (2013): 131–147.

31. The value of this distinction dissolves with the formation of satratoxins by the poison fire coral because satratoxins are also produced by molds. The presence of the same toxins in mushrooms and molds is explained by the fact that some fungi that produce mushrooms have a second identity as molds. A brief explanation of this complex feature of fungal life cycles is provided by Sarah C. Watkinson, Lynne Boddy, and Nicholas P. Money, *The Fungi*, 3rd ed. (Amsterdam: Academic Press, 2016), 20–21. The fire coral is an ascomycete mushroom, which is a closer relation to morels than gilled mushrooms and boletes, and its asexual stages are

classified as species of *Trichoderma*: Gary J. Samuels and D. J. Lodge, "Three Species of *Hypocrea* with Stipitate Stromata and *Trichoderma* Anamorphs," *Mycologia* 88, no. 2 (1996): 302–315.

32. Caroline De Costa, "St Anthony's Fire and Living Ligatures: A Short History of Ergometrine," *Lancet* 359, no. 9319 (2002): 1768–1770.

33. Yan Liu, *Healing with Poisons: Potent Medicines in Medieval China* (Seattle: University of Washington Press, 2021).

34. Carolyn A. Young, Christopher L. Schardl, Daniel G. Panaccione, Simona Florea, Johanna E. Takach, Nikki D. Charlton, Neil Moore, et al., "Genetics, Genomics and Evolution of Ergot Alkaloid Diversity," *Toxins (Basel)* 7, no. 4 (2015): 1273–1302.

35. Laurinda S. Dixon, "Bosch's 'St. Anthony Triptych'—An Apothecary's Apotheosis," *Art Journal* 44 (2014): 119–131.

36. Linnda R. Caporael, "Ergotism: The Satan Loosed in Salem?," *Science* 192, no. 4234 (1976): 21–26.

37. P. Salway and W. Dell, "Plague at Athens," *Greece and Rome* 2, no. 2 (1955): 62–69; Mary K. Matossian, *Poisons of the Past: Molds, Epidemics and History* (New Haven, CT: Yale University Press, 1989).

38. A. J. Holladay and J. C. F. Poole, "Thucydides and the Plague of Athens," *The Classical Quarterly* 29 (1979): 282–300; Jane Bellemore, Ian M. Plant, and Lynne M. Cunningham, "Plague of Athens—Fungal Poison?," *Journal of the History of Medicine and Allied Sciences* 49, no. 4 (1994): 521–545. The German pharmacologist and toxicologist Rudolf Kobert (1854–1918) believed that the symptoms of the plague might have been caused by a combination of a smallpox outbreak in a population already weakened by ergotism.

39. Abraham Z. Joffe, "Alimentary Toxic Aleukia," in *Algal and Fungal Toxins*, ed. Solomon Kadis, Alex Ciegler, and Samuel J. Ajl (New York: Academic Press, 1971), 139–189; and "*Fusarium poae* and *F. sporotrichioides* as Principal Causal Agents of Alimentary Toxic Aleukia," in *Mycotoxic Fungi, Mycotoxins, Mycotoxicoses: An Encyclopedic Handbook*, vol. 3, *Mycotoxicoses of Man and Plants: Mycotoxin Control and Regulatory Practices*, ed. Thomas D. Wyllie and Lawrence G. Morehouse (New York: Marcel Dekker, 1978), 21–86.

40. Outbreaks of ergotism in the twentieth century included a spate of cases among Jewish immigrants from Central Europe living in Manchester in 1927, and 250 poisonings in the town of Pont St. Esprit in southern France in the 1950s. Although ergotism fits many of the facts of the mass psychosis that afflicted the residents of Pont St. Esprit, mercury contamination of bread flour is one of several alternative explanations. Other eruptions occurred in India, and ergotism continues to flare up in Ethiopia: Sarah Belser-Ehrlich, Ashley Harper, John Hussey, and Robert Hallock, "Human and Cattle Ergotism since 1900: Symptoms, Outbreaks, and Regulations," *Toxicology and Industrial Health* 29, no. 4 (2013): 307–316.

41. Noreddine Benkerroum, "Chronic and Acute Toxicities of Aflatoxins: Mechanisms of Action," *International Journal of Environmental Research and Public Health* 17, no. 2 (2020): 423; Stephanie Kraft, Lisa Buchenauer, and Tobias Polte, "Mold, Mycotoxins and a Dysregulated Immune System: A Combination of Concern?," *International Journal of Molecular Sciences* 22, no. 22 (2021): 12269.

42. J. W. Bennett and M. Klich, "Mycotoxins," *Clinical Microbiology Reviews* 16, no. 3 (2003): 497–516.

43. Yun Yun Gong, Sinead Watson, and Michael N. Routledge, "Aflatoxin Exposure and Associated Human Health Effects, a Review of Epidemiological Studies," *Food Safety (Japan)* 4, no. 1 (2016): 14–27.

44. Robert J. Lee, Alan D. Workman, Ryan M. Carey, Bei Chen, Philip L. Rosen, Laurel Doghramji, Nithin D. Adappa, et al., "Fungal Aflatoxins Reduce Respiratory Mucosal Ciliary Function," *Scientific Reports* 6 (2016): 33221.

45. Dr. Harriet Burge, a distinguished professor in the Harvard School of Public Health, demonstrated the improbability of significant inhalational exposure to mycotoxins in mold-damaged homes by calculating the number of spores inhaled per hour: Harriet A. Burge, "Fungi: Toxic Killers or Unavoidable Nuisances?," *Annals of Allergy, Asthma, and Immunology* 87 (2001): 52–56.

46. Nicholas P. Money, *Carpet Monsters and Killer Spores: A Natural History of Toxic Mold* (New York: Oxford University Press, 2004).

47. Joan W. Bennett, "The Fungi That Ate My House," *Science* 349 (2015): 1018; Arati A. Inamdar, Shannon Morath, and Joan W. Bennett, "Fungal Volatile Organic Compounds: More Than Just a Funky Smell?," *Annual Review of Microbiology* 74, no. 1 (2020): 101–116.

48. Nandhitha Venkatesh and Nancy P. Keller, "Mycotoxins in Conversation with Bacteria and Fungi," *Frontiers in Microbiology* 10 (2019): 403; Daniel G. Panaccione, "Origins and Significance of Ergot Alkaloid Diversity in Fungi," *FEMS Microbiology Letters* 251, no. 1 (2005): 9–17.

49. The potency of mycotoxins has not been lost on military strategists, and molds are undoubtedly part of secret bioweapons research. Mary K. Klassen-Fischer, "Fungi as Bioweapons," *Clinics in Laboratory Medicine* 26, no. 2 (2006): 387–395; Edyta Janik-Karpińska, Michał Ceremunga, Joanna Saluk-Bijak, and Michał Bijak, "Biological Toxins as the Potential Tools for Bioterrorism," *International Journal of Molecular Sciences* 20 (2019): 1181.

50. Nicholas P. Money, *The Rise of Yeast: How the Sugar Fungus Shaped Civilization* (Oxford: Oxford University Press, 2018).

CHAPTER NINE

1. Robert Alter, *The Hebrew Bible*, vol. 2, *Prophets* (New York: Norton, 2019), Ezekiel 1:15–17, pp. 1054–1055; Jacques M. Chevalier, *A Postmodern Revelation: Signs of Astrology and the Apocalypse* (Toronto: University of Toronto Press, 1997), 223–263; Shawn Z. Aster, "Ezekiel's Adaptation of Mesopotamian *Melammu*," *Die Welt des Orients* 45, no. 1 (2015): 10–21.

2. Flavie Waters, Jan D. Blom, Thien T. Dang-Vu, Allan J. Cheyne, Ben Alderson-Day, Peter Woodruff, and Daniel Collerton, "What Is the Link between Hallucinations, Dreams, and Hypnagogic-Hypnopompic Experiences?," *Schizophrenia Bulletin* 42, no. 5 (2016): 1098–1109; Rainer Kraehenmann, "Dreams and Psychedelics: Neurophenomenological Comparison and Therapeutic Implications," *Current Neuropharmacology* 15, no. 7 (2017): 1032–1042; Camila Sanz, Federico Zamberlan, Earth Erowid, Fire Erowid, and Enzo Tagliazucchi, "The Experience Elicited by Hallucinogens Presents the Highest Similarity to Dreaming within a Large Database of Psychoactive Substance Reports," *Frontiers in Neuroscience* 12 (2018): 7; Benjamin Baird, Sergio A. Mota-Rolim, and Martin Dresler, "The Cognitive Neuroscience of Lucid Dreaming," *Neuroscience and Biobehavioral Reviews* 100 (2019): 305–323. Lucid dreaming refers to dreams in

which we become aware that we are dreaming as the action takes place, but there does not seem to be a clear distinction between this experience and very vivid dreams like my fantasy of the swirling cosmos.

3. Psilocin slips through cell membranes more easily than serotonin and binds with receptor proteins inside nerve cells. Serotonin stays on the outside. This difference may explain some of the longer-term effects of the mushroom alkaloid on the nervous system: Maxemiliano V. Vargas, Lee E. Dunlap, Chunyang Dong, Samuel J. Carter, Robert J. Tombari, Shekib A. Jami, Lindsay P. Cameron, et al., "Psychedelics Promote Neuroplasticity through the Activation of Intracellular 5-HT2A Receptors," *Science* 379 (2023): 700–706.

4. Jiawei Zhang, "Basic Neural Units of the Brain: Neurons, Synapses and Action Potential," May 30, 2019, arXiv:1906.01703.

5. The description of the brain as a computer is apt, as long as we recognize the limitations of this metaphor. Unlike a digital computer, the brain is an analog device that processes information by gathering data from multiple sources to produce approximate answers or personal representations rather than the precise and unvarying output of computers. The digital description is more useful at the cellular level because each of the nerve cells in the brain is limited to transmitting or blocking an incoming electrical signal: Romaine Brette, "Brains as Computers: Metaphor, Analogy, Theory or Fact?," *Frontiers in Ecology and Evolution* 10 (2022): 878729; Blake A. Richards and Timothy P. Lillicrap, "The Brain-Computer Metaphor Debate Is Useless: A Matter of Semantics," *Frontiers of Computer Science* 4 (2022): 810358. It is also noteworthy that the three-pound ball of jelly in the skull draws no more energy than a lightbulb, whereas the supercomputer lives in an air-conditioned vault and consumes more electricity than a small city.

6. Drummond E.-W. McCulloch, Gitte M. Knudsen, Frederick S. Barrett, Manoj K. Doss, Robin L. Carhart-Harris, Fernando E. Rosas, Gustavo Deco, et al., "Psychedelic Resting-State Neuroimaging: A Review and Perspective on Balancing Replication and Novel Analyses," *Neuroscience and Biobehavioral Reviews* 138 (2022): 104689.

7. N. L. Mason, K. P. C. Kuypers, F. Müller, J. Reckweg, D. H. T. Tse, S. W. Toennes, N. R. P. W. Hutten, et al., "Me, Myself, Bye: Regional Alterations in Glutamate and the Experience of Ego Dissolution with Psilocybin," *Neuropsychopharmacology* 45 (2020): 2003–2011.

8. Lea J. Mertens, Matthew B. Wall, Leor Roseman, Lysia Demetriou, David J. Nutt, and Robin L. Carhart-Harris, "Therapeutic Mechanisms of Psilocybin: Changes in Amygdala and Prefrontal Functional Connectivity during Emotional Processing after Psilocybin for Treatment-Resistant Depression," *Journal of Psychopharmacology* 34, no. 2 (2020): 167–180. The amygdala, or amygdala nuclei, are paired clusters of neurons buried in the brain that are involved in processing memories, making decisions, and controlling fear, aggression, and anxiety.

9. Nina Schimmel, Joost J. Breeksema, Sanne Y. Smith-Apeldoorn, Jolien Veraart, Wim van den Brink, and Robert A. Schoevers, "Psychedelics for the Treatment of Depression, Anxiety, and Existential Distress in Patients with a Terminal Illness: A Systematic Review," *Psychopharmacology (Berlin)* 239, no. 1 (2022): 15–33.

10. Gabrielle I. Agin-Liebes, Tara Malone, Matthew M. Yalch, Sarah E. Mennenga, K. Linnae Ponté, Jeffrey Guss, Anthony P. Bossis, et al., "Long-Term Follow-Up of Psilocybin-Assisted Psychotherapy for Psychiatric and Existential Distress in Patients with Life-Threatening Cancer," *Journal of Psychopharmacology* 34, no. 2 (2020): 155–166.

11. Erwin Krediet, Tijmen Bostoen, Joost Breeksema, Annette van Schagen, Torsten Passie, and Eric Vermetten, "Reviewing the Potential of Psychedelics for the Treatment of PTSD," *International Journal of Neuropsychopharmacology* 23, no. 6 (2020): 385–400; Michael P. Bogenschutz, Stephen Ross, Snehal Bhatt, Tara Baron, Alyssa A. Forcehimes, Eugene Laska, Sarah E. Mennenga, et al., "Percentage of Heavy Drinking Days Following Psilocybin-Assisted Psychotherapy vs Placebo in the Treatment of Adult Patients with Alcohol Use Disorder: A Randomized Clinical Trial," *JAMA Psychiatry* (2022), doi:10.1001/jamapsychiatry.2022.2096; Meg J. Spriggs, Hannah M. Douglass, Rebecca J. Park, Tim Read, Jennifer L. Danby, Frederico J. C. de Magalhães, Kirsty L. Alderton, et al., "Study Protocol for 'Psilocybin as a Treatment for Anorexia Nervosa: A Pilot Study,'" *Frontiers in Psychiatry* 12 (2021): 735523.

12. Richard E. Daws, Christopher Timmermann, Bruna Giribaldi, James D. Sexton, Matthew B. Wall, David Erritzoe, Loer Roseman, et al., "Increased Global Integration in the Brain after Psilocybin Therapy for Depression," *Nature Medicine* 28, no. 4 (2022): 844–851; Ling-Xiao Shao, Clara Liao, Ian Gregg, Pasha A. Davoudian, Neil K. Savalia, Kristin Delagarza, and Alex C. Kwan, "Psilocybin Induces Rapid and Persistent Growth of Dendritic Spines in Frontal Cortex In Vivo," *Neuron* 109, no. 16 (2021): 2535–2544.

13. Sean McClintock, "Why Investors Are Turning toward Psychedelic Healthcare Companies," *Fortune*, September 4, 2021; Yeji J. Lee, "What to Know about the Booming Psychedelics Industry Where Companies Are Racing to Turn Magic Mushrooms and MDMA into Approved Medicines," *Insider*, June 30, 2022; Michelle Lhooq, "With Magic Mushrooms, Small Businesses Lead, Hoping Laws Will Follow," *Bloomberg Businessweek*, July 21, 2022.

14. "Oregon Psilocybin Services Section Summary of Measure 109: Listening Session December 13–15, 2021," Oregon Health Authority, December 2021, https://www.oregon.gov/oha/PH/PREVENTIONWELLNESS/Documents/M109-Summary-2021-Dec.pdf.

15. Andrew Selsky, "Oregon Voters Face 2 Drug Measures on November Ballot," *AP News*, November 4, 2020.

16. Theresa M. Carbonaro, Matthew P. Bradstreet, Frederick S. Barrett, Katherine A. MacLean, Robert Jesse, Matthew W. Johnson, and Roland R. Griffiths, "Survey Study of Challenging Experiences after Ingesting Psilocybin Mushrooms: Acute and Enduring Positive and Negative Consequences," *Journal of Psychopharmacology* 30, no. 12 (2016): 1268–1278.

17. Andy Letcher, *Shroom: A Cultural History of the Magic Mushroom* (London: Faber and Faber, 2006).

18. O. T. Oss and O. N. Oeric, *Psilocybin: Magic Mushroom Grower's Guide* (Berkeley, CA: And/Or Press, 1976). Otos is derived from the Greek word meaning insatiate (never satisfied) and oneiric is an adjective that refers to dreams. The coauthors were pseudonyms for McKenna, who wrote the foreword for his book under his real name.

19. N. Milne, P. Thomsen, N. Mølgaard Knudsen, P. Rubaszka, M. Kristensen, and I. Borodina, "Metabolic Engineering of *Saccharomyces cerevisiae* for the *de Novo* Production of Psilocybin and Related Tryptamine Derivatives," *Metabolic Engineering* 60 (2020): 25–36; William J. Gibbons, Madeline G. McKinney, Philip J. O'Dell, Brooke A. Bollinger, and J. Andrew Jones, "Homebrewed Psilocybin: Can New Routes for Pharmaceutical Psilocybin Production Enable Recreational Use?," *Bioengineered* 12, no. 1 (2021): 8863–8871.

20. Janis Fricke, Felix Blei, and Dirk Hoffmeister, "Enzymatic Synthesis of Psilocybin," *Angewandte Chemie International Edition* 56, no. 40 (2017): 12352–12355; R. C. Van Court, M. S. Wiseman, K. W. Meyer, D. J. Ballhorn, K. R. Amses, J. C. Slot, B. T. M. Dentinger, et al., "Diversity, Biology, and History of Psilocybin-Containing Fungi: Suggestions for Research and Technological Development," *Fungal Biology* 126, no. 4 (2022): 308–319.

21. Hannah T. Reynolds, Vinod Vijayakumar, Emile Gluck-Thaler, Hailee Brynn Korotkin, Patrick Brandon Matheny, and Jason C. Slot, "Horizontal Gene Cluster Transfer Increased Hallucinogenic Mushroom Diversity," *Evolution Letters* 2, no. 2 (2018): 88–101.

22. Kevin McKernan, Liam Kane, Yvonne Helbert, Lei Zhang, Nathan Houde, and Stephen McLaughlin, "A Whole Genome Atlas of 81 *Psilocybe* Genomes as a Resource for Psilocybin Production," *F1000Research* 10 (2021): 961.

23. M. Hibicke and C. D. Nichols, "Validation of the Forced Swim Test in *Drosophila*, and Its Use to Demonstrate Psilocybin Has Long-Lasting Antidepressant-Like Effects in Flies," *Scientific Reports* 12 (2022): 10019. Psilocybin appears to boost the optimism of fruit flies immersed in water without any means of escape. This experiment is a scaled-down version of an unpleasant laboratory test in which rodents are dropped into glass cylinders filled with water to induce feelings of hopelessness. The animals struggle to climb out of the water before giving up and resorting to paddling to stay afloat. Animals attempt to escape this hopeless situation longer and harder if they are treated with antidepressants, which seems to parallel the development and relief of depression in humans. Like mice and rats, fruit flies fed with psilocybin struggle longer than controls, who give up the ghost after a few seconds. The details of the experiment show that the fruit flies respond to immersion in water with periods of immobility interspersed with activity and a shortening of the immobile periods under the influence of psilocybin. Fungus gnats have not been tortured in the same way, but their brains are similar to fruit flies, and our brains are constructed from the same cellular hardware as those of insects.

24. Although most spores are dispersed from gilled mushrooms by wind, insects that visit fruit bodies consume spores and carry them in their digestive systems when they fly away. These spores are deposited in the feces of the insects, which provides nutritional support for the growth of the young mycelia when they germinate. The insect attraction model for psilocybin is supported by the presence of the highest levels of the compound in the caps of these mushrooms: Klára Gotvaldová, Kateřina Hájková, Jan Borovička, Radek Jurok, Petra Cihlářová, and Martin Kuchař, "Stability of Psilocybin and Its Four Analogs in the Biomass of the Psychotropic Mushroom *Psilocybe cubensis*," *Drug Testing and Analysis* 13 (2021): 439–446.

25. Ali R. Awan, Jaclyn M. Winter, Daniel Turner, William M. Shaw, Laura M. Suz, Alexander J. Bradshaw, Tom Ellis, et al., "Convergent Evolution of Psilocybin Biosynthesis by Psychedelic Mushrooms," *bioRxiv* (2018), https://doi.org/10.1101/374199.

26. Brian Lovett, Raymond J. St. Leger, and Henrik H. de Fine, "Going Gentle into That Pathogen-Induced Goodnight," *Journal of Invertebrate Pathology* 174 (2020): 107398. "Pathogen-Induced Good Night" would be Thomasian and make more grammatical sense.

27. Greg R. Boyce, Emile Gluck-Thaler, Jason C. Slot, Jason E. Stajich, William J. Davis, Tim Y. James, John R. Cooley, et al., "Psychoactive Plant- and Mushroom-Associated Alkaloids from Two Behavior Modifying Cicada Pathogens," *Fungal Ecology* 41 (2019): 147–164.

28. Claudius Lenz, Jonas Wick, Daniel Braga, María García-Altares, Gerald Lackner, Christian Hertweck, Markus Gressler, et al., "Injury-Triggered Blueing Reactions of *Psilocybe* "Magic" Mushrooms," *Angewandte Chemie International Edition* 59, no. 4 (2020): 1450–1454.

29. Quentin Carboué and Michel Lopez, "*Amanita muscaria*: Ecology, Chemistry, Myths," *Encyclopedia* 1 (2021): 905–914.

30. In English, fly was a familiar term for a demon in the sixteenth century. Reginald Scot, *The Discoverie of Witchcraft* (London: William Brome, 1584), refers to "a flie, otherwise called a divell or familiar" (III, xv, p. 65), and "*Beelzebub*, which signifieth the lord of the flies, bicause he taketh everie simple thing in his web" (xix, p. 518). Pieces of the mushroom soaked or boiled in milk attract flies that are poisoned by ibotenic acid, which is the precursor or prodrug of muscimol: Mateja Lumpert and Samo Kreft, "Catching Flies with *Amanita muscaria*: Traditional Recipes from Slovenia and Their Efficacy in the Extraction of Ibotenic Acid," *Journal of Ethnopharmacology* 187 (2016): 1–8.

31. Jan D. Blom, "Alice in Wonderland Syndrome: A Systematic Review," *Neurology Clinical Practice* 6, no. 3 (2016): 259–270.

32. L. Alison McInnes, Jimmy J. Qian, Rishab S. Gargeya, Charles DeBattista, and Boris D. Heifets, "A Retrospective Analysis of Ketamine Intravenous Therapy for Depression in Real-World Care Settings," *Journal of Affective Disorders* 301 (2022): 486–495.

33. Francesca I. Rampolli, Premiila Kamler, Claudio C. Carlino, and Francesca Bedussi, "The Deceptive Mushroom: Accidental *Amanita muscaria* Poisoning," *European Journal of Case Reports in Internal Medicine* 8, no. 3 (2021): 002212. The same toxins are responsible for poisonings by the panther cap, *Amanita pantherina*: Leszek Satora, Dorota Pach, Krysztof Ciszowski, and Lidia Winnik, "Panther Cap *Amanita pantherina* Poisoning Case Report and Review," *Toxicon* 47, no. 5 (2006): 605–607.

34. The literature on ethnomycology is vast. If readers are unfamiliar with this subject and are interested in exploring it further, a simple web search will lead to a wealth of online essays, books, and podcasts on the topic. The following paper also provides a helpful overview of the field: Giorgio Samorini, "The Oldest Archeological Data Evidencing the Relationship of *Homo sapiens* with Psychoactive Plants: A Worldwide Overview," *Journal of Psychedelic Studies* 3, no. 2 (2019): 63–80.

35. Alter, *Hebrew Bible*, vol. 2, Ezekiel 28:13–14, p. 1136.

36. Robert Graves, "Mushrooms, Food of the Gods," *The Atlantic*, August 1957, https://www.math.uci.edu/~vbaranov/nicetexts/eng/mushrooms.html.

37. R. Gordon Wasson, *Soma: Divine Mushroom of Immortality* (New York: Harcourt, Brace & World, 1969); Kevin Feeney, "Revisiting Wasson's Soma: Exploring the Effects of Preparation on the Chemistry of *Amanita Muscaria*," *Journal of Psychoactive Drugs* 42, no. 4 (2010): 499–506.

38. John M. Allegro, *The Sacred Mushroom and the Cross: A Study of the Nature and Origins of Christianity within the Fertility Cults of the Ancient Near East* (London: Hodder & Stoughton, 1970).

39. C. F. Evans, "The Scholars and the World of God," *The Times* (London), November 11, 1971. Wasson was appalled by Allegro's shoddy scholarship, writing, "I think that he jumped to unwarranted conclusions on scanty evidence. And when you make such blunders as attributing

the Hebrew language, the Greek language, to Sumerian—that is unacceptable to any linguist. The Sumerian language is parent to no language and no one knows where it came from." This critique is quoted from the following compilation: Jan Irvin, "The Defamation of Allegro," in Jan Irvin and Andrew Rutajit, *Astrotheology and Shamanism* (San Diego: The Book Tree, 2005), 51–58, http://www.johnallegro.org/the-defamation-of-allegro-by-jan-irvin-excerpted-from -astrotheology-shamanism/.

40. Jerry B. Brown and Julie M. Brown, *The Psychedelic Gospels: The Secret History of Hallucinogens in Christianity* (Rochester, VT: Park Street Press, 2016); and "Entheogens in Christian Art: Wasson, Allegro, and the Psychedelic Gospels," *Journal of Psychedelic Studies* 3, no. 2 (2019): 142–163.

41. R. R. Griffiths, W. A. Richards, U. McCann, and R. Jesse, "Psilocybin Can Occasion Mystical-Type Experiences Having Substantial and Sustained Personal Meaning and Spiritual Significance," *Psychopharmacology* 187 (2006): 268–283; R. R. Griffiths, W. A. Richards, M. W. Johnson, U. McCann, and R. Jesse, "Mystical-Type Experiences Occasioned by Psilocybin Mediate the Attribution of Personal Meaning and Spiritual Significance 14 Months Later," *Journal of Psychopharmacology* 22, no. 6 (2008): 621–632.

42. Roland R. Griffiths, Ethan S. Hurwitz, Alan K. Davis, Matthew W. Johnson, and Robert Jesse, "Survey of Subjective 'God Encounter Experiences': Comparisons among Naturally Occurring Experiences and Those Occasioned by the Classic Psychedelics Psilocybin, LSD, Ayahuasca, or DMT," *PLoS ONE* 14, no. 4 (2016): e0214377.

43. For a provocative and objective article about the interpretation of mystical experiences produced by drug use, see Huston Smith, "Do Drugs Have Religious Import?," *Journal of Philosophy* 61, no. 18 (1964): 517–530.

44. Aldous Huxley, *The Doors of Perception & Heaven and Hell* (New York: Harper, 2009). Huxley's description of the flower arrangement appears on pp. 16–17. Huxley took his title from William Blake, *The Marriage of Heaven and Hell* (undated poem written in the 1790s), and Jim Morrison, the name of his band. Without the assistance of mushrooms, Blake wrote, "If the doors of perception were cleansed every thing would appear to man as it is, Infinite."

45. Aldous Huxley, *Brave New World* (London: Chatto and Windus, 1932), and *Brave New World Revisited* (New York: Harper, 1958).

46. Robin L. Carhart-Harris, Robert Leech, Peter J. Hellyer, Murray Shanahan, Amanda Feilding, Enzo Tagliazucchi, Dante R. Chialvo, et al., "The Entropic Brain: A Theory of Conscious States Informed by Neuroimaging Research with Psychedelic Drugs," *Frontiers in Human Neuroscience* 8 (2014): 20; Rubén Herzog, Pedro A. M. Mediano, Fernando E. Rosas, Robin Carhart-Harris, Yonatan S. Prl, Enzo Tagliazucchi, and Rodrigo Cofre, "A Mechanistic Model of the Neural Entropy Increase Elicited by Psychedelic Drugs," *Scientific Reports* 10 (2020): 17725.

47. Steven D. Hollon, Paul W. Andrews, Daisy R. Singla, Marta M. Maslej, and Benoit H. Mulsant, "Evolutionary Theory and the Treatment of Depression: It Is All About the Squids and the Sea Bass," *Behavior Research and Therapy* 143 (2021): 103849.

48. Robert Burton, *The Anatomy of Melancholy* (Oxford: John Litchfield and James Short, for Henry Cripps, 1621), Part II, Sect. 3. The quote derives from Horace's *Odes*, I.24.

49. Chris Paling, *A Very Nice Rejection Letter: Diary of a Novelist* (London: Constable, 2021), 151.

50. W. Steven Gilbert, *The Life and Work of Dennis Potter* (Woodstock, NY: Overlook Press, 1998), 294.

51. Microdosing has become a popular approach to achieving the supposed creative benefits of the drug without losing practical contact with the immediate tasks of the day: Federico Cavanna, Stephanie Muller, Laura A. de la Fuente, Federico Zamberlan, Matías Palmucci, Lucie Janeckova, Martin Kuchar, et al., "Microdosing with Psilocybin Mushrooms: A Double-Blind Placebo-Controlled Study," *Translational Psychiatry* 12 (2022): 307. Unfortunately, this study found no evidence that microdosing increased feelings of well-being, creativity, or cognitive function beyond a placebo effect.

CHAPTER TEN

1. Stephen R. Kane, Zhexing Li, Eric T. Wolf, Colby Ostberg, and Michelle L. Hill, "Eccentricity Driven Climate Effects in the Kepler-1649 System," *Astronomical Journal* 161, no. 1 (2020): 31. Kepler 1649c is three quadrillion kilometers from Earth, and it would take six million years for peopled or unpeopled probes to get there.

2. Yinon M. Bar-On, Rob Phillips, and Ron Milo, "The Biomass Distribution on Earth," *Proceedings of the National Academy of Sciences USA* 115, no. 25 (2018): 6506–6511. Plants make up 80 percent of the weight of the biosphere; 2 percent of the biomass lives in fungi, and animals contribute less than 1 percent to the sum of biology. Land plants make up most of the billions of tons of botany, and one-third of the weight of plants is in their roots, where they form mycorrhizas with fungi.

3. Billions of elms and American chestnuts were wiped out in the twentieth century, and eucalyptus trees and pines are plagued by rusts today. Roderick J. Fensham and Julian Radford-Smith, "Unprecedented Extinction of Tree Species by Fungal Disease," *Biological Conservation* 261 (2021): 109276; Erin Shanahan, Kathryn M. Irvine, David Thoma, Siri Wilmoth, Andrew Ray, Kristin Legg, and Henry Shovic, "Whitebark Pine Mortality Related to White Pine Blister Rust, Mountain Pine Beetle Outbreak, and Water Availability," *Ecosphere* 7, no. 12 (2016): e01610. These pandemic diseases are spread by global commerce and exacerbated by climate change.

4. N. C. Johnson, J. H. Graham, and F. A. Smith, "Functioning of Mycorrhizal Associations along the Mutualism-Parasitism Continuum," *New Phytologist* 135, no. 4 (1997): 575–586; Nancy-Collins Johnson and James H. Graham, "The Continuum Concept Remains a Useful Framework for Studying Mycorrhizal Functioning," *Plant and Soil* 363 (2013): 411–419; Marc-André Selosse, Laure Schneider-Maunoury, and Florent Martos, "Time to Re-Think Fungal Ecology? Fungal Ecological Niches Are Often Prejudged," *New Phytologist* 217 (2018): 968–972. Even the supposedly saintly mycorrhizal fungi can stray from their benevolence toward plants by becoming antagonistic toward their hosts and behaving as parasites. Truffles are ectomycorrhizal with oaks and hazels and create clear patches around their hosts called *brûlés* by parasitizing weeds and grasses that compete for soil nutrients: I. Plattner and I. R. Hall, "Parasitism of Non-Host Plants by the Mycorrhizal Fungus *Tuber melanosporum*," *Mycological Research* 99, no. 11 (1995): 1367–1370. Matsutake species seem particularly catholic, feeding as mutualists, as parasites, and as saprotrophs on dead roots: Wang Yun, Ian R. Hall, and Lynley A. Evans, "Ectomycorrhizal Fungi with Edible Fruiting Bodies 1. *Tricholoma matsutake* and Related Fungi," *Economic Botany*

51, no. 3 (1997): 311–327; Lin-Min Vaario, Taina Pennanen, Tytti Sarjala, Eira-Maija Savonen, and Jussi Heinonsalo, "Ectomycorrhization of *Tricholoma matsutake* and Two Major Conifers in Finland—An Assessment of In Vitro Mycorrhiza Formation," *Mycorrhiza* 20, no. 7 (2010): 511–518; Wang Yun, "Matsutake: A Natural Biofertilizer?," in *Handbook of Microbial Fertilizers*, ed. M. K. Rai (Binghamton, NY: Food Products Press, 2006), 497–541.

5. Suzanne W. Simard, David A. Perry, Melanie D. Jones, David D. Myrold, Daniel M. Durall, and Randy Molina, "Net Transfer of Carbon between Ectomycorrhizal Tree Species in the Field," *Nature* 388 (1997): 579–582. The importance of the fungi in nutrient transfer between trees was questioned when this classic study was published and remains controversial: David Robinson and Alastair Fitter, "The Magnitude and Control of Carbon Transfer between Plants Linked by a Common Mycorrhizal Network," *Journal of Experimental Botany* 50, no. 330 (1999): 9–13; Monika A. Gorzelak, Benjamin H. Ellert, and Leho Tedersoo, "Mycorrhizas Transfer Carbon in a Mature Mixed Forest," *Molecular Ecology* 29 (2020): 2315–2317; Justine Karst, Melanie D. Jones, and Jason D. Hoeksema, "Positive Citation Bias and Overinterpreted Results Lead to Misinformation on Common Mycorrhizal Networks in Forests," *Nature Ecology and Evolution* (2023), https://doi.org/10.1038/s41559-023-01986-1.

6. Thomas I. Wilkes, "Arbuscular Mycorrhizal Fungi in Agriculture," *Encyclopedia* 1 (2021): 1132–1154; Manjula Novindarajulu, Philip E. Pfeffer, Hairu Jin, Jehad Abubaker, David D. Douds, James W. Allen, Heike Bücking, et al., "Nitrogen Transfer in the Arbuscular Mycorrhizal Symbiosis," *Nature* 435 (2005): 819–823; Joanne Leigh, Angela Hodge, and Alastair H. Fitter, "Arbuscular Mycorrhizal Fungi Can Transfer Substantial Amounts of Nitrogen to Their Host Plant from Organic Material," *New Phytologist* 181, no. 1 (2009): 199–207; Sally E. Smith, Iver Jakobsen, Mette Grønlund, and F. Andrew Smith, "Roles of Arbuscular Mycorrhizas in Plant Phosphorus Nutrition: Interactions between Pathways of Phosphorus Uptake in Arbuscular Mycorrhizal Roots Have Important Implications for Understanding and Manipulating Plant Phosphorus Acquisition," *Plant Physiology* 156, no. 3 (2011): 1050–1057; Kevin Garcia and Sabine D. Zimmermann, "The Role of Mycorrhizal Associations in Plant Potassium Nutrition," *Frontiers in Plant Science* 5 (2014): 337.

7. Ruairidh J. H. Sawers, M. Rosario Ramírez-Flores, Víctor Olalde-Portugal, and Uta Paszkowski, "The Impact of Domestication and Crop Improvement on Arbuscular Mycorrhizal Symbiosis in Cereals: Insights from Genetics and Genomics," *New Phytologist* 220, no. 4 (2018): 1135–1140; Jeremiah A. Henning, Evan Weiher, Yali D. Lee, Deborah Freund, Artur Stefanski, and Stephen P. Bentivenga, "Mycorrhizal Fungal Spore Community Structure in a Manipulated Prairie," *Restoration Ecology* 26 (2018): 124–133.

8. Laura A. Bolte, Arnau V. Vila, Floris Imhann, Valerie Collij, Ranko Gacesa, Vera Peters, Cisca Wijmenga, et al., "Long-Term Dietary Patterns Are Associated with Pro-Inflammatory and Anti-Inflammatory Features of the Gut Microbiome," *Gut* 70, no. 7 (2021): 1287–1298; Bernard Srour, Melissa C. Kordahi, Erica Bonazzi, Mélanie Deschasaux-Tanguy, Mathilde Touvier, and Benoit Chassaing, "Ultra-Processed Foods and Human Health: From Epidemiological Evidence to Mechanistic Insights," *Lancet Gastroenterology and Hepatology* 7 (2022): 1128–1140. The explicit effect of a fast-food diet on the gut fungi is inferred from studies on mice (see chapter 5): Tahliyah S. Mims, Qusai Abdallah, Justin D. Stewart, Sydney P. Watts, Catrina T. White, Thomas V. Rousselle, Ankush Gosain, et al., "The Gut Mycobiome of Healthy Mice Is

Shaped by the Environment and Correlates with Metabolic Outcomes in Response to Diet," *Communications Biology* 4, no. 1 (2021): 281; Jata Shankar, "Food Habit Associated Mycobiota Composition and Their Impact on Human Health," *Frontiers in Nutrition* 8 (2021): 773577.

9. Karin Hage-Ahmed, Kathrin Rosner, and Siegred Steinkellner, "Arbuscular Mycorrhizal Fungi and Their Response to Pesticides," *Pest Management* 75, no. 3 (2019): 583–590; Anna Edlinger, Gina Garland, Kyle Hartman, Samiran Banerjee, Florine Degrune, Pablo García-Palacios, Sara Hallin, et al., "Agricultural Management and Pesticide Use Reduce the Functioning of Beneficial Plant Symbionts," *Nature Ecology and Evolution* 6 (2022): 1145–1154; Gavin Duley and Emanuele Boselli, "Mutual Plant-Fungi Symbiosis Compromised by Fungicide Use," *Communications Biology* 5 (2022): 1069.

10. Megan H. Ryan and James Graham, "Little Evidence That Farmers Should Consider Abundance or Diversity of Arbuscular Mycorrhizal Fungi When Managing Crops," *New Phytologist* 220, no. 4 (2018): 1092–1107; Matthias C. Rillig, Carlos A. Aguilar-Trigueros, Tessa Camenzind, Timothy R. Cavagnaro, Florine Degrune, Pierre Hohmann, Daniel R. Lammel, et al., "Why Farmers Should Manage the Arbuscular Mycorrhizal Symbiosis," *New Phytologist* 222, no. 3 (2019): 1171–1175.

11. Zahangir Kabir, "Tillage or No-Tillage: Impact on Mycorrhizae," *Canadian Journal of Plant Science* 85, no. 1 (2015): 23–29; Xingli Lu, Xingneng Lu, and Yuncheng Liao, "Effect of Tillage Treatment on the Diversity of Soil Arbuscular Mycorrhizal Fungal and Soil Aggregate-Associated Carbon Content," *Frontiers in Microbiology* 9 (2018): 2986; Chen Zhu, Ning Ling, Junjie Guo, Min Wang, Shiwei Guo, and Qirong Shen, "Impacts of Fertilization Regimes on Arbuscular Mycorrhizal Fungal (AMF) Community Composition Were Correlated with Organic Matter Composition in Maize Rhizosphere Soil," *Frontiers in Microbiology* 7 (2016): 1840.

12. Inês Rocha, Isabel Duarte, Ying Ma, Pablo Souza-Alonso, Aleš Látr, Miroslav Vosátka, Helena Freitas, et al., "Seed Coating with Arbuscular Mycorrhizal Fungi for Improved Field Production of Chickpea," *Agronomy* 9 (2019): 471.

13. M. Eric Benbow, Philip S. Barton, Michael D. Ulyshen, James C. Beasley, Travis L. DeVault, Michael S. Strickland, Jeffery K. Tomberlin, et al., "Necrobiome Framework for Bridging Decomposition Ecology of Autotrophically and Heterotrophically Derived Organic Matter," *Ecological Monographs* 89, no. 1 (2019): e01331; Peter G. Kennedy and François Maillard, "Knowns and Unknowns of the Soil Fungal Necrobiome," *Trends in Microbiology* 31, no. 2 (2023): 173–180.

14. J. J. C. Sidrim, R. E. Moreira Filho, R. A. Cordeiro, M. F. G. Rocha, E. P. Caetano, A. J. Monteiro, and R. S. N. Brilhante, "Fungal Microbiota Dynamics as a Postmortem Investigation Tool: Focus on *Aspergillus*, *Penicillium* and *Candida* Species," *Journal of Applied Microbiology* 108 (2010): 1751–1756; Xiaoliang Fu, Juanjuan Guo, Dmitrijs Finkelbergs, Jing He, Lagabaiyila Zha, Yadong Guo, and Jifeng Cai, "Fungal Succession during Mammalian Cadaver Decomposition and Potential Forensic Implications," *Scientific Reports* 9 (2019): 12907.

15. Zohreh Shariatinia, "Heidegger's Ideas about Death," *Pacific Science Review B: Humanities and Social Sciences* 1, no. 2 (2015): 92–97. This short paper by an Iranian scholar covers the essentials of Heidegger's thinking on death without a hint of philosophical jargon.

16. Katie Rogers, "Mushroom Suits, Biodegradable Urns and Death's Green Frontier," *New York Times*, April 22, 2016.

17. Piratical metaphors are very helpful for explaining biological facts: Nicholas P. Money and Mark W. F. Fischer, "What Is the Weight of a Single Amoeba and Why Does It Matter?," *American Biology Teacher* 83, no. 9 (2021): 571–574.

18. Thomas Terberger, Mikhail Zhilin, and Svetlana Savchenko, "The Shigir Idol in the Context of Early Art in Eurasia," *Quaternary International* 573 (2021): 1–3.

19. Joëlle Dupont, Claire Jacquet, Bruno Dennetière, Sandrine Lacoste, Faisl Bousta, Geneviève Orial, Corinne Cruaud, et al., "Invasion of the French Paleolithic Painted Cave of Lascaux by Members of the *Fusarium solani* Species Complex," *Mycologia* 99, no. 4 (2007): 526–533.

20. Pedro Martin-Sanchez, Alena Novakova, Fabiola Bastian, Claude Alabouvette, and Cesareo Saiz-Jimenez, "Two New Species of the Genus *Ochroconis*, *O. lascauxensis* and *O. anomala* Isolated from Black Stains in Lascaux Cave, France," *Fungal Biology* 116 (2012): 574–589.

21. Laura Zucconi, Fabiana Canini, Daniela Isola, and Giulia Caneva, "Fungi Affecting Wall Paintings of Historical Value: A Worldwide Meta-Analysis of Their Detected Diversity," *Applied Sciences* 12 (2022): 2988.

22. Nahid Akhtar and M. Amin-Ul Mannan, "Mycoremediation: Expunging Environmental Pollutants," *Biotechnology Reports (Amsterdam)* 26 (2020): e00452; A. Arun and M. Eyini, "Comparative Studies on Lignin and Polycyclic Aromatic Hydrocarbons Degradation by Basidiomycetes Fungi," *Bioresource Technology* 102, no. 17 (2011): 8063–8070.

23. Roc Tkavc, Vera Y. Matrosova, Olga E. Grichenko, Cene Gostinčar, Robert P. Volpe, Polina Klimenkova, Elena K. Gaidamakova, et al., "Prospects for Fungal Bioremediation of Acidic Radioactive Waste Sites: Characterization and Genome Sequence of *Rhodotorula taiwanensis* MD1149," *Frontiers in Microbiology* 8 (2018): 2528. The fungus in this study is a yeast rather than a filamentous fungus, which is unusually tolerant to gamma radiation.

24. Anna Lowenhaupt Tsing, *The Mushroom at the End of the World: On the Possibility of Life in Capitalist Ruins* (Princeton, NJ: Princeton University Press, 2015); Alison Pouliot, *The Allure of the Fungi* (Clayton South, Australia: CSIRO Publishing, 2018).

25. A. Johnson, "Blackfoot Indian Utilization of the Flora of the Northwestern Great Plains," *Economic Botany* 24 (1970): 301–324; William R. Burk, "Puffball Usages among North American Indians," *Journal of Ethnobiology* 3 (1983): 55–62.

26. The study of the fungi began in 1729 with the publication of Micheli's *Nova Plantarum Genera* (see chapter 7, note 32). Corrado Nai and Vera Meyer, "The Beauty and the Morbid: Fungi as Source of Inspiration in Contemporary Art," *Fungal Biology and Biotechnology* 3 (2016): 10; Regine Rapp, "On Mycohuman Performances: Fungi in Current Artistic Research," *Fungal Biology and Biotechnology* 6 (2019): 22.

27. Ofer Grunwald, Ety Harish, and Nir Osherov, "Development of Novel Forms of Fungal Art Using *Aspergillus nidulans*," *Journal of Fungi* 7, no. 12 (2021): 1018.

28. Emily Farra, "You Aren't Tripping: Fungi Are Taking Over Fashion," *Vogue*, April 2, 2021.

29. Patricia Kaishian and Hasmik Djoulakian, "The Science Underground: Mycology as a Queer Discipline," *Catalyst: Feminism, Theory, Technoscience* 6, no. 2 (2020): 1–26.

30. Nicholas P. Money, "Obituary: Cecil Terence Ingold (1905–2010)," *Nature* 465 (2010): 1025.

31. Martin Grube, Ester Gaya, Håvard Kauserud, Adrian M. Smith, Simon Avery, Sara J. Fernstad, Lucia Muggia, et al., "The Next Generation Fungal Diversity Researcher," *Fungal Biology Reviews* 31, no. 3 (2017): 124–130.

32. Nicholas P. Money, "Hyphal and Mycelial Consciousness: The Concept of the Fungal Mind," *Fungal Biology* 125, no. 4 (2021): 257–259; Kristin Aleklett and Lynne Boddy, "Fungal Behaviour: A New Frontier in Behavioural Ecology," *Trends in Ecology and Evolution* 36, no. 9 (2021): 787–796. Each cubic centimeter or milliliter of brain tissue contains sixty-eight million neurons, which is similar to the maximum number of hyphae that can be packed into the same volume of soil.

33. Mohammad Mahdi Dehshibi and Andrew Adamatzky, "Electrical Activity of Fungi: Spikes Detection and Complexity Analysis," *Biosystems* 203 (2021): 104373; Andrew Adamatzky, "Language of Fungi Derived from Their Electrical Spiking Activity," *Royal Society Open Science* 9, no. 4 (2022): 211926.

34. Rhawn G. Joseph, Richard Armstrong, Xinli Wei, Carl Gibson, Olivier Planchon, David Duvall, Ashraf M. T. Elewa, et al., "Fungi on Mars? Evidence of Growth and Behavior from Sequential Images," *Journal of Cosmology* 29, no. 4 (2021): 480–550.

35. DNA profiles from human blood samples can be recovered after they are burned and reach a temperature of 1,000 degrees Celsius: A. Klein, O. Krebs, A. Gehl, J. Morgner, L. Reeger, C. Augustin, and C. Edler, "Detection of Blood and DNA Traces after Thermal Exposure," *International Journal of Legal Medicine* 132, no. 4 (2018): 1025–1033.

36. Gerald R. Taylor, Mary R. Henney, and Walter L. Ellis, "Changes in the Fungal Autoflora of Apollo Astronauts," *Applied Microbiology* 26, no. 5 (1973): 804–813.

37. Adriana Blachowicz, Snehit Mhatre, Nitin K. Singh, Jason M. Wood, Ceth W. Parker, Cynthia Ly, Daniel Butler, et al., "The Isolation and Characterization of Rare Mycobiome Associated with Spacecraft Assembly Cleanrooms," *Frontiers in Microbiology* 13 (2022): 777133.

38. Aleksandra Checinska, Alexander J. Probst, Parag Vaishampayan, James R. White, Deepika Kumar, Victor G. Stepanov, George E. Fox, et al., "Microbiomes of the Dust Particles Collected from the International Space Station and Spacecraft Assembly Facilities," *Microbiome* 3 (2015): 50.

39. Takashi Sugita, Takashi Yamazaki, Otomi Cho, Satoshi Furukawa, and Chiaki Mukai, "The Skin Mycobiome of an Astronaut during a 1-Year Stay on the International Space Station," *Medical Mycology* 59, no. 1 (2021): 106–109.

40. Donatella Tesei, Anna Jewczynko, Anne M. Lynch, and Camilla Urbaniak, "Understanding the Complexities and Changes in the Astronaut Microbiome for Successful Long-Duration Space Missions," *Life* 12 (2022): 495.

APPENDIX

1. Jie Tang, Iliyan D. Iliev, Jordan Brown, David M. Underhill, and Vincent A. Funari, "Mycobiome: Approaches to Analysis of Intestinal Fungi," *Journal of Immunological Methods* 421 (2015): 112–121; Robert Edgar, "Taxonomy Annotation and Guide Tree Errors in 16S rRNA Databases," *PeerJ* 6 (2018): e5030.

2. Amanda K. Dupuy, Marika S. David, Lu Li, Thomas N. Heider, Jason D. Peterson, Elizabeth A. Montano, Anna Dongari-Bagtzoglou, et al., "Redefining the Human Oral Mycobiome with Improved Practices in Amplicon-Based Taxonomy: Discovery of *Malassezia* as a Prominent Commensal," *PLoS ONE* 9, no. 3 (2014): e90899; Mallory J. Suhr and Heather E.

Hallen-Adams, "The Human Gut Mycobiome: Pitfalls and Potentials—A Mycologist's Perspective," *Mycologia* 107, no. 6 (2015): 1057–1073.

3. Analysis of the fungi found in the sputum of asthma patients in Wandsworth, in south London, identified some rather unlikely species: Hugo C. van Woerden, Clive Gregory, Richard Brown, Julian R. Marchesi, Bastiaan Hoogendoorn, and Ian P. Matthews, "Differences in Fungi Present in Induced Sputum Samples from Asthma Patients and Non-Atopic Controls: A Community Based Case Control Study," *BMC Infectious Diseases* 13 (2013): 69. Hugo Cornelis and his team from the Cardiff University School of Medicine reported that one of the species that was prevalent in the lungs of asthma patients and absent in non-asthmatic controls was *Termitomyces clypeatus*. This fungus produces a large mushroom that was discovered in the 1920s in a bamboo thicket in the Democratic Republic of the Congo, then the colony of the Belgian Congo, where it was fruiting from an abandoned termite mound. The mycelium of this mushroom is farmed by termites, who consume scraps of wood and plant leaves and defecate onto a spongy "comb" that is colonized by the fungus. Most of the plant matter eaten by the insects is indigestible, like the fiber in our diet, which is where the mushroom comes in. By decomposing the fiber, the fungal mycelium becomes enriched with protein and fat that serves as the perfect food for the termites. The description of this species was not published until 1951: Roger Heim, "Les *Termitomyces* du Congo Belge Recueillis par Madame M. Goossens-Fontana," *Bulletin du Jardin Botanique de l'État Bruxelles* 21, no. 3, 4 (1951): 205–222. *Termitomyces clypeatus* has a wide geographical distribution and is sold in local markets in Cameroon and Nigeria as a flavorful mushroom with purported medicinal properties: Oumar Mahamat, Njouonkou André-Ledoux, Tume Chrisopher, Abamukong Adeline Mbifu, and Kamanyi Albert, "Assessment of Antimicrobial and Immunomodulatory Activities of Termite Associated Fungi, *Termitomyces clypeatus* R. Heim (Lyophyllaceae, Basidiomycota)," *Clinical Phytoscience* 4 (2018): 28. Wandsworth is a land of many splendors, but termite mounds are scarce. The presence of this mushroom in the sputum samples alleged by Van Worden was cited uncritically by Laura Tipton, Elodie Ghedin, and Alison Morris, "The Lung Mycobiome in the Next-Generation Sequencing Era," *Virulence* 8, no. 3 (2017): 334–341, which illustrates how errors can persist in the literature when investigators have minimal knowledge of the organisms that show up in the DNA analyses. Besides this African toadstool, the extensive list of species identified from the lungs of asthmatic Londoners in the Van Worden study included a wood-rotter from forests in the Southern Hemisphere, a fungus that grows inside eucalyptus trees in South Africa, and, strangest of all, a little mushroom that fruits beneath the water in cold Argentinian lakes. The wood-decay fungus supposedly found in this study is *Grifola sordulenta*; *Lasiodiplodia gonubiensis* is the South African endophyte; and *Gloiocephala aquatica* is the aquatic mushroom. There are many other species listed in this report that are also unlikely to be floating in the fragrant air of London.

4. Through this encounter I had, inadvertently, changed places with Professor Heinz Wolff (1928–2017), a well-known British academic, who stopped by my demonstration of sperm release in ferns at a science fair in Oxford, England, in the late 1970s, and peered into my microscope. I was very pleased with myself for figuring out how to coerce explosions of swimming spermatozoids from tiny fern plantlets. Wolff asked me in his German accent why I had bothered to do this, saying, "But vot experiment hev you performed here?" (Think Peter Sellers as Dr. Strangelove.) It was a good question. My project was observational and only minimally

experimental, and I admitted as much. He was not impressed and left my table shaking his head at what he appeared to perceive as a brief conversation with a sixteen-year-old imbecile. My science teacher was more supportive and cursed Wolff, after he was beyond earshot, with a shocking train of expletives. The prize winners that day were boys from a private school (ours was the local "comprehensive") who had developed a nuclear reactor or something similarly impressive.

5. Nicholas P. Money, "Against the Naming of Fungi," *Fungal Biology* 117 (2013): 463–465. In this publication, I wrote, "For 250 years mycologists have tried to reconcile fungal diversity with the Linnean fantasy of a divine order throughout nature that included unambiguous species. This effort has failed and today's taxonomy rests on an unstable philosophical foundation." We lack a robust definition of a fungal species, which has led to treating some populations of fungi that are only distantly related as members of the same species, and, in other cases, assigning more than one name to fungi that others regard as single species.

6. Petr Kralik and Matteo PandRicchi, "A Basic Guide to Real Time PCR in Microbial Diagnostics: Definitions, Parameters, and Everything," *Frontiers in Microbiology* 8 (2017): 108; M. N. Zakaria, M. Furuta, T. Takeshita, Y. Shibata, R. Sundari, N. Eshima, T. Ninomiya, et al., "Oral Mycobiome in Community-Dwelling Elderly and Its Relation to Oral and General Health Conditions," *Oral Diseases* 23, no. 7 (2017): 973–982.

Illustrations

Chapter 7   Beautiful fruit bodies of bird's nest fungi that are the source of antibiotics called cyathanes. ©2013 Insil Choi.

Chapter 8   St. Anthony of Egypt, whose relics became associated with the miraculous cure of ergotism in the Middle Ages. The victim of ergotism in this sixteenth-century woodcut is suffering from the *ignis sacer* or holy fire, which was a burning sensation in the extremities caused by vasoconstriction, and has lost one of his legs below the knee to gangrene. *Source:* World History Archive / Alamy Stock Photo.

Chapter 9   Fruit bodies of a species of *Psilocybe* with hallucinogenic properties. ©2022 Insil Choi.

Chapter 10   Mycorrhizal symbiosis between the roots of a tree and the mycelia of mushrooms that are fruiting at the surface of the soil. ©2022 Insil Choi.

Index